illumination engineering

for energy efficient
luminous environments

illumination engineering
for energy efficient luminous environments

RONALD N. HELMS

University of Colorado
Boulder, Colorado

Prentice-Hall, Inc.
Englewood Cliffs, NJ 07632

Library of Congress Cataloging in Publication Data

Helms, Ronald N.
    Illumination engineering for energy efficient
luminous environments.

    Includes bibliographies and index.
    1. Electric lighting. 2. Lighting. 3. Electric
power—Conservation. I. Title.
TK4175.H45   621.32'2   79-14866
ISBN 0-13-450809-2

Editorial/production supervision and interior design by Steven Bobker.
Page layout by Diane Heckler-Koromhas.
Cover design by Jorge Hernandez .
Manufacturing buyer: Gordon Osbourne

10   9   8   7   6   5   4   3   2   1

Printed in the United States of America

Prentice-Hall International, Inc., *London*
Prentice-Hall of Australia Pty. Limited, *Sydney*
Prentice-Hall of Canada, Ltd., *Toronto*
Prentice-Hall of India Private Limited, *New Delhi*
Prentice-Hall of Japan, Inc., *Tokyo*
Prentice-Hall of Southeast Asia Pte. Ltd., *Singapore*
Whitehall Books Limited, *Wellington, New Zealand*

To

Lawrence Nelson Helms
(1911-1977)

This book is dedicated to my father, Nelson Helms. My dad, who worked for 40 years in a steel mill, wanted a better life for his boys. "If you want to end up working a swing shift all your life like your dad, don't get a college education. If you want a college education, I'll do everything possible to help." My dad passed away in August, 1977 of lung cancer at the young age of 66 years.

To
Robert Nelson, Laura Elaine, Cathline
Carole, James Mitchell, and Margo Baron Helms.

contents

preface

This book has its beginnings in 1965 when I came to the University of Colorado to "develop a program in Illumination–Electrical Systems in Architectural Engineering," according to Professor Robert E. Rathburn. It was Bob's dream to make architectural engineering the training grounds for all of the consulting engineering professions (structural, mechanical, and electrical) as well as construction engineering and management. The book is an outgrowth of lecture notes developed by myself and used in Illumination I–Lighting Fundamentals and Illumination II–Applications in Lighting. These two courses are a part of the Illumination–Electrical option curriculum in Architectural Engineering at the University of Colorado.

Many excellent books on illuminating engineering and such related fields as optics, light sources, and color have been written over the years since Thomas Edison received his "patent on a new, high resistance carbon filament vacuum lamp" (Chapter 4) in 1880. These books run the gamut from lighting in its pure art form to highly complex mathematical treatment of the subject. This book attempts to cover the aesthetic as well as the technical aspects of illumination engineering.

I have spent my summers since 1965 working for electrical consulting engineers in the Denver area, and more recently as a selfemployed private lighting consultant. I have always felt strongly that my students could gain valuable experience from my practical encounters in the field. This touch with reality has given me insight into the shortcomings of most illumination engineers and lighting designers. The illuminating engineer may be well grounded in the technical aspects (calculations) while the lighting designer has an understanding of the artistic and aesthetic aspects of lighting. Unfortunately, both are usually lacking in one or more of the important fundamental concepts. The key to being a complete illuminating engineer (or lighting designer) is to be both an artist and an engineer. The illuminating engineer must know all aspects of lighting, from the pure art form to the highly complex mathematical technique. Knowing these fundamental concepts, one must know how to mix art and engineering to make a cohesive, balanced lighting solution. The task of providing well balanced lighting solutions has been made more difficult with the advent of the energy crisis and pending energy shortages (Chapter 8). The visibility of lighting makes it a natural target for those uneducated and undereducated in lighting fundamentals. The solution to energy efficient luminous environments is a part of all chapters in this book. Important fundamental concepts are found in each chapter. All of these concepts must weigh heavily on the balance that must be struck between visibility, visual comfort, productivity, performance, aesthetics, cost, and energy conservation.

The most important Chapters are 1 through 4 since these chapters deal with fundamental concepts of importance to all fields and applications. Most errors in lighting design can be traced back to a lack of understanding of subjects covered in the first four chapters. If the book is to be used as a textbook, a good fundamentals course can be built around Chapter 1-6. A second, follow-up Course in applications could be based on Chapters 7 through Chapter 10. Specialized courses such as a Roadway Lighting Seminar might make use of Chapters 1 to 5, part of Chapter 7 (Point-by-Point), and Chapter 9. Any specialized seminars utilizing this book should always include at least Chapter 1 through Chapter 4.

I wish to acknowledge two of my professors from my undergraduate days at the University of Illinois. Professor Steve Tang, my mentor and advisor, who by his example and deeds, planted the seeds that would one day bring me into the teaching profession. Professor Tang was asked by a student why he had given up a high paid, secure job as chief engineer in a large architectural and engineering firm. Professor Tang's reply was "the profession has been good to me and I have learned a lot. It was time for me to give something back to the profession. Too many people gain valuable knowledge and experience that they end up taking to the grave. I want to pass some of my experience on to future generations while I am young and capable."

Professor Roland Pierce first spurred my interest in illuminating engineering. He provided additional guidance at the graduate level through individual tutorial sessions and special projects in illumination. He was instrumental in bringing the University of Colorado and me together in 1965. When I first began teaching, Professor Pierce came out to the University of Colorado to help me establish my first photometric lab and experiments. Many thanks to both Professor Tang and Professor Pierce for their guidance, encouragement, and untiring dedication to their students–particularly me.

In 1969, I returned to school after five years of teaching to obtain my Ph.D. in Biophysics. My two-year program could not have been possible without the guidance and assistance of Dr. H. Richard Blackwell. I thank Dr. Blackwell for his patience and understanding of the time pressures. The wisdom and guidance of Glenn Fry and the inspirational teaching of Jack King made my educational experience at Ohio State a time to remember.

I also wish to thank Mr. John Kaufman, Technical Director of the Illuminating Engineering Society of North America for his technical review of this book. Mr. Lewis O. Harvey, Jr. was kind enough to review the sections on vision and color. Thanks to both of these men for their guidance and review of the book.

Ronald N. Helms

1

light,
eye,
vision

PHYSICS OF LIGHT

The two basic theories associated with lighting research are the quantum and the electromagnetic theories (a form of wave theory). Planck's quantum theory (1905) states that each quantum is present in the wave as a separate entity so that the radiation takes on the appearance of a stream of bullets called photons. This was a return to Newton's corpuscular theory, except the corpuscles are now called photons. The quantum theory was not widely accepted because at the time it could not be used to explain interference and defraction phenomena.

Quantum (corpuscular) — $\epsilon = h\nu$

Wave Theory — $\lambda = \dfrac{c}{\nu}$

Period T = Time for cycle

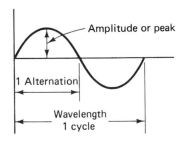

Figure 1-1 Wave Terminology

ν — frequency, hertz
ϵ — Quantum (am't of energy)
h — Planck's Constant—
  6.6256×10^{-27} erg/sec.
λ — Wavelength
c — Speed of light in a vacuum—
  approximately 186,000 miles/
  sec or 3×10^8 m/sec.

Duality Theory

In 1924, de Broglie developed a hypothesis that combined the wave theory and corpuscular (or particle) theory. Depending on the wavelength (Figure 1-1), either the wave aspect or the particle aspect may be dominant. For example, radio waves are strictly wavelike, whereas cosmic rays behave like particles.

Electromagnetic Spectrum

The electromagnetic theory provides the best explanation of the radiant energy characteristics most frequently used by the illuminating engineer. A graphical representation of radiant energy (Figure 1-2) is called a **spectrum.** Different forms of radiant energy are laid out along a continuum from cosmic rays to electricity. The electromagnetic (or radiant energy) spectrum ranges from 3.937×10^{-13} inch (in.) or 10^{-5} nanometer (nm) for cosmic rays to 3100 miles (4.98×10^{15} nm) for 60-hertz (Hz) electric current.

The dividing line between two forms of radiant energy is not as sharp as Figure 1-2 might indicate but rather there is a gradual transition from one to another. Because of the reciprocal relationship between wavelength and frequency ($\lambda = c/\nu$), radiant energy can be expressed in terms of its wavelength or frequency along the continuum. Classically, **wavelength** has been used to describe all forms of radiant energy left of the dividing line (Figure 1-2) between infrared and radar. Radiant energy to the right of this dividing line is described in terms of its **frequency.** This dividing line between infrared and radar is probably due to the physical size of the numbers when expressed in terms of wavelength or frequency. For example, electricity is most commonly designated in terms of its frequency, such as 60 Hz. However, it could also be designated in terms of its wavelength, which is approximately 10^{15} nm. That is a number followed by 15 zeros, which is not a convenient number to work with.

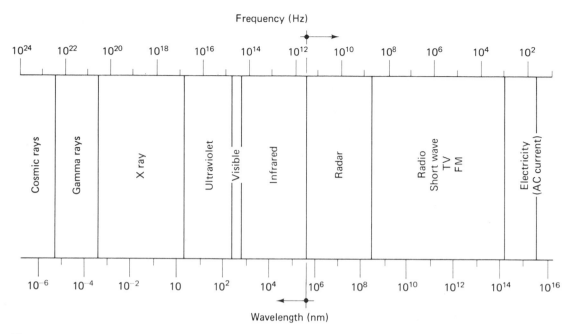

Figure 1-2. Electromagnetic Spectrum

Visible Spectrum

In illumination we are concerned with a small portion of the electromagnetic spectrum called the **visible spectrum.** The visible spectrum derives its name from the fact that any energy produced in this narrow band will produce the sensation of vision when it stimulates a normal human eye. The visible spectrum has the following wavelength range:

From violet: 0.38×10^{-4} cm $= 0.38 \times 10^{-6}$ m $= 0.38 \, \mu m$ (micrometers)
$= 3800$ Å (angstroms) $= 15$ millionths of an inch
$= 380$ nm (nanometers)

To red: 0.76×10^{-4} cm $= 0.76 \times 10^{-6}$ m $= 0.76 \, \mu m = 7600$ Å
$= 760$ nm $= 30$ millionths of an inch

Each band or part of the visible spectrum will produce a different color sensation when the eye is light adapted (see Table 1-1).

Purple exists in the spectrum; however, it occurs because of a red cone–blue cone interaction. That is, purple is an artifact of the visual system and not a pure spectral color.

The phenomena of color sensation that involves the interaction of visible radiant energy and the human visual system is a very important concept. If a "normal" human visual system were stimulated with pure 500-nm (single

TABLE 1-1

Color Versus Wavelength

Color	Wavelength (nm)
Red	760-630
Orange	630-590
Yellow	590-560
Green	560-490
Blue	490-440
Indigo	440-420
Violet	420-380
Purple	Not a pure spectral color

wavelength) radiant energy, the light-adapted subject would respond that the light is green. If the wavelength were changed to pure 585-nm radiant energy, the response would be that the light is yellow. Therefore, the wavelength composition of light is important to the sensation of color perception.

ANATOMY OF THE EYE

Structure of the Eye

The eye or globe is a spherically shaped precision instrument. The globe is made up of three concentric layers (Figure 1-3) referred to as **tunics.**

Outer Tunic

The **outer tunic or fibrous tunic** is divided into two distinctly different parts. The anterior sixth is the **cornea;** the posterior five-sixths is the **sclera.** The cornea is the most important refractive median in the eye. It is the outer transparent membrane that refracts or bends the radiant energy so that it can be focused on the retina. The sclera is a tough outer membrane that acts to maintain the shape of the globe.

Middle Tunic

The **middle tunic, vascular tunic, or uvea** is subdivided into three parts: choroid, ciliary body, and iris. The most important function of the **choroid** is to provide nourishment for the outer vascular layers of the retina (layers 1 through 4; see Figure 1-4). The **ciliary body** is important in changing the shape of the lens in **accommodation.** The ciliary body also produces the aqueous humor that fills the outer chamber of the eye. The aqueous humor is a watery fluid that helps to refract and filter light entering the eye. The **iris** adjusts the amount of light that enters the eye and helps to increase the depth of focus.

The **ciliary zonules** are dense condensed fibrals that run from the ciliary body in two sheets, where they fuse with the capsule of the lens. The lens is involved with changing the dioptics of the eye in accommodation; the lens and iris

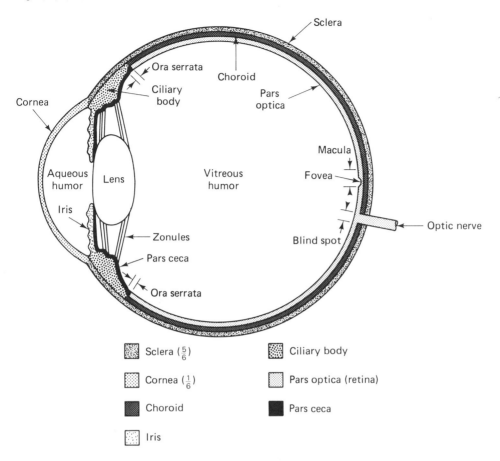

Figure 1-3. Tunics of the Eye

are the major optical components in the functioning of the eye. As an object is brought close to the eye, muscles in the ciliary body act on the zonules and cause the lens to bulge, which brings the object into focus. At the same time the iris opens or closes depending upon the amount of light available. As the object is moved away from the eye, the muscles relax and cause the lens to flatten and the eye to focus. The process of lens focusing is called **accommodation.** The magnitude of accommodation decreases with age, causing blurred near vision. This decrease in the magnitude of accommodation is believed to be due to a hardening of the lens substance. This condition is known as **presbyopia** and begins at about the age of 40.

The eye is said to be functioning in distant vision range when the object is more than 20 feet (ft) from the eye. Since the muscles that control accommodation and convergence are in a relaxed state during distant vision, the eye will have less strain and fatigue.

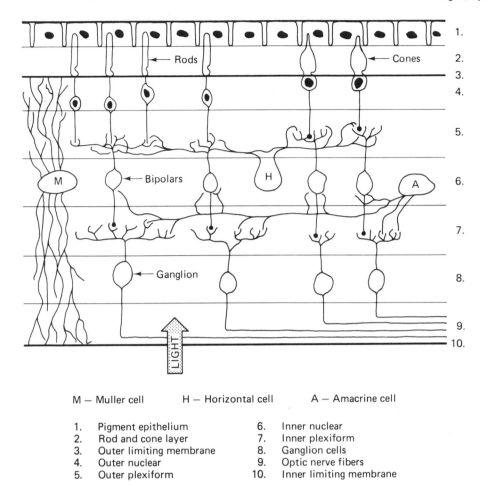

M — Muller cell   H — Horizontal cell   A — Amacrine cell

1.	Pigment epithelium	6.	Inner nuclear
2.	Rod and cone layer	7.	Inner plexiform
3.	Outer limiting membrane	8.	Ganglion cells
4.	Outer nuclear	9.	Optic nerve fibers
5.	Outer plexiform	10.	Inner limiting membrane

Figure 1-4. Layers of the Retina

Innermost Tunic

The **innermost tunic or nervous tunic** is divided into the pars optica (retina) and the pars ceca. The nervous tunic is an outgrowth of the brain called the **diencephalon.** The pars ceca is the anterior (front) most portion of this tunic. It is represented as a thin layer on the innermost aspect of the iris and ciliary body. The pars ceca ends at the **ora serrata.** The ora serrata (see Figure 1-3) is the transition area between the choroid and ciliary body at the level of the vascular tunic, and the transition area between the pars ceca and pars optica at the level of the nervous tunic. There are no visual cells in the ora serrata. The pars optica or retina proper (retina) is the major posterior (back) light-sensitive portion of the nervous tunic. Two important specialized areas are associated with the pars optica (retina): the **optic disc** and the **macula.**

The optic disc (white spot or blind spot) is the area where the optic nerve leaves the retina (see Figure 1-3). There are no visual cells at the blind spot. The blind spot appears as a white spot approximately 1.5 millimeters (mm) in diameter when viewed through the front of the eye. The macula (macula lutea or yellow spot) (see Figures 1-3 and 1-6) is a depression about 1.5 mm in diameter (5.2° external) and is located 3.5 mm temporal and 0.5 mm inferior to the optic disc. The bottommost part of the depression is called the fovea centralis or **fovea.** It is an area of about 0.4 mm in diameter (1.4° external). The fovea is the most acute visual area on the retina. The muscles of the eye and the refractive media act to focus light on the fovea. The fovea is the point of clearest vision and best color response.

Retina

The retina, which is the inner lining of the eye, is the receiving or light-sensitive portion. The retina contains delicate light-sensitive nerve fiber endings called **cones** and **rods.**

The retina consists of 10 layers of cells and processes. Figure 1-4 is a schematic cross section of the retina indicating the layers and the names associated with each layer.

Cones

The cones are effective for "daylight" vision only. The maximum concentration of cones (Figure 1-5) is in the fovea, and they decrease in number as they move outward from the fovea. The fovea is inactive under very dim light (less than 0.01 footlambert; fL). The footlambert (see Chap. 3) is a unit of luminous energy leaving a surface and arriving at the retina. The rods must take over the visual process during low levels of illumination. When the eye is receiving light at levels about approximately 1 fL, the system is said to be operating under **photopic** or pure cone vision.

$$\text{cones} \left(\frac{1}{100} \text{ fL to } \infty \right) \quad \text{pure cone } (\sim 1 \text{ fL to } \infty)$$

While moving from a dark environment into a very light environment, the visual system experiences a change in sensitivity. This phenomena is called **light adaptation.** Light adaptation involves primarily the cone system, and usually takes less than a minute.

Rods

The rods function primarily during "night" vision. There are no rods in the fovea. The rods increase in number (Figure 1-5) from just outside the fovea outward. The rods are highly sensitive to light and motion, which is responsible for sharp peripheral vision. A bundle of rods (many thousand) is served by one nerve ending. These multiple hookups result in very poor visual acuity. There is no color response with the rod systems. Rods produce a black and white

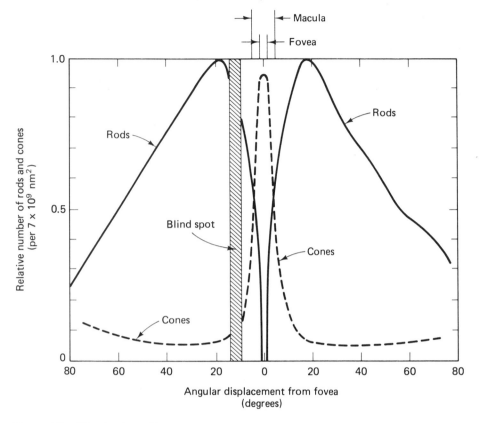

Figure 1-5. Distribution of Rods and Cones

response, which is actually a change in luminance. Moving from a very light environment into a dark environment results in a change in sensitivity of the visual system due to **dark adaptation.** Dark adaptation is not completely achieved until about 1 hour (h) after light is removed. The initial rapid phase of dark adaptation is due to low-level cone action, while the remaining portion of dark adaptation is due to slower rod action. **Scotopic** or rod vision occurs when the eye has been dark adapted and only the rods are functioning.

$$\text{rods} \ (10^{-6} \ \text{fL to} \sim 1 \ \text{fL})$$

Rods and Cones

When both rods and cones are functioning, the system is said to be in the **mesopic** range. The mesopic range is from $\frac{1}{100}$ fL to approximately 1 fL. The upper limit is dependent on the size, position, and time of exposure of the rod system. This is the range most commonly encountered during nighttime activities such as driving.

Distribution of Cones and Rods

There are 6 to 7 million cones and 75 to 150 million rods distributed across the surface of the retina. Figure 1-5 shows the relative distribution of the cones and rods. The break in the curves (parallel lines) represents the location of the optic disc or blind spot. The largest concentration of both cones and rods is in the region of the macula. Figure 1-6 identifies the area and external angle subtended by areas in the region of the fovea. The external angle represents the projection of an area on the retina out into the external world outside the eye. The external angle is referred to as the **visual angle** (see Visual Acuity). This angle can be used to determine the physical size of objects at various distances that can be seen by each region (Figure 1-6) of the retina.

Figure 1-6. Areas Within the Macula Region

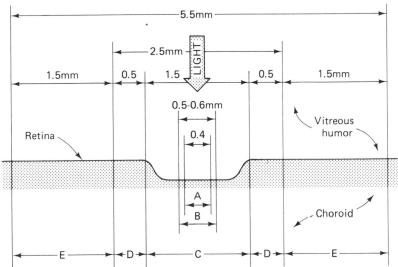

Region	Diameter		External angle, degrees (object space)
	Microns (μ)	mm	
A. Fovea	400	0.4	1.4°
B. Rod free area	500 – 600	0.5 to 0.6	1.7 – 2.0°
C. Mucula	1500	1.5	5.2°
D. Parafovea (band approx. 500 μ wide)	2500	2.5	8.6°
E. Perifovea (band approx. 1500 μ wide)	5500	5.5	19 – 20°

VISUAL PATHWAY

Radiant energy in the visible spectrum passes through the eye and is focused on the retina. The light stimulus causes a photochemical reaction in the rods and cones. This reaction causes nerve impulses to be generated in ganglion cells, which are then transmitted through ganglion cell fibers to the region of the optic disc. The impulse continues along the fibrous processes (Figure 1-7) from the globe along the optic nerve where half of the fibers decussate (cross over) in a structure called the **optic chiasma.** The optic fibers leaving the chiasma change in name and make up. Thereafter, the fibers are called the **optic tract.** The impulses travel through the fibers of the optic tract and terminate in the **lateral geniculate body** (LGB), which is located in the thalamus of the brain. The cells of the LGB give rise to new fibers, which form the **optic radiation** fibers that extend out and back through the brain to the cortex (surface of the brain) in a region of the brain called the **occipital lobes** (Figure 1-8). The area of the cortex that receives the optic radiations surrounds the calcarine fissure and is called the striate area, striate cortex, visual area, or Brodmann's Area 17. The gross area of vision in the brain is called the **occipital cortex.**

Figure 1-7. Visual Pathway

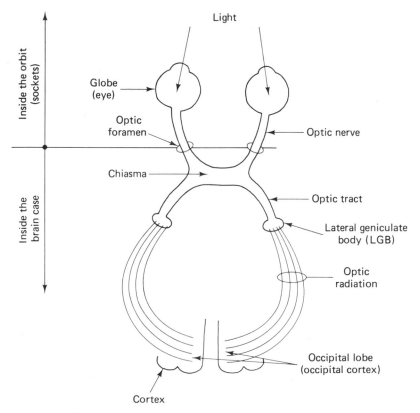

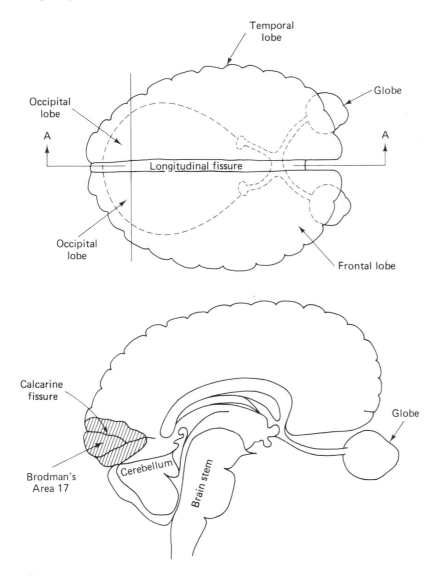

Figure 1-8. Brain and Occipital Cortex

SENSITIVITY OF THE VISUAL SYSTEM

The luminous efficiency of radiant energy is dependent on the ability of various receiving and measuring devices. The spectral response characteristics of the human eye vary between individuals so that it is not feasible for any one individual to act as a standard observer. Therefore, experiments had to be conducted on thousands of subjects to determine the average response characteristics of the human visual system.

Threshold Curve (Figure 1-9)

The spectral sensitivity of the human eye to the entire visible spectrum can be established experimentally. The scotopic (dim-light) experiment was initially conducted by Hecht and Williams in 1922. In 1932, Gibson and Tyndall conducted a bright-light experiment to obtain the photopic curve. These experiments involved measuring the threshold response to stimulus at various wavelengths. Threshold can be thought of as the yes/no point of vision, that is, the point at which some critical detail just disappears, the "I see it/I don't see it" point of vision.

The interval between the absolute threshold of visibility (scotopic curve) and the initial appearance of hue for a given homogeneous wavelength (photopic curve) is called the **photochromatic interval.** This interval is greatest at the shorter wavelengths (blue) and near zero at the longer wavelengths (red). Mesopic vision occurs at intermediate levels of luminance, where rods and cones are believed to work together.

Sensitivity Curve (Figure 1-10)

By definition, sensitivity has a reciprocal relationship to threshold. That is,

$$\text{sensitivity} = \frac{1}{T_\lambda}$$

This results in the threshold curve being inverted and allows one to speak in terms of sensitivity, which is a more meaningful term to most individuals.

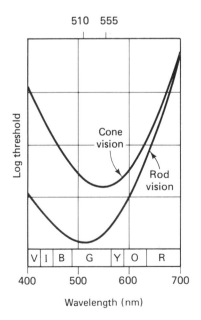

Figure 1-9. Threshold

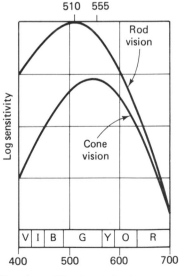

Figure 1-10. Sensitivity

12

Four major facts can be derived from sensitivity curves:

1. Neither the rods nor cones are uniformly sensitive across the visible spectrum.

2. The region of maximal sensitivity is 555 nm for cone vision and 510 nm for rod vision.

3. The rod function lies above the cone function, indicating that throughout most of the spectrum the rods require less energy for vision (have a greater sensitivity) than the cones.

4. The rods and cones are equally sensitive to radiant energy in the long-wavelength (red) end of the spectrum.

Visual stimulation at scotopic levels does not produce color vision anywhere along the spectrum. Color vision is possible only with light levels of sufficient magnitude to activate the cone system. When only the rods are functioning, all wavelengths are seen as a series of lighter or darker grays. Weak light is visible, but hue is absent.

Relative Sensitivity (Figure 1-11)

The previous curves are more commonly presented in the form of photopic and scotopic relative luminosity curves (relative spectral sensitivity curve or standard observer curve):

$$V_\lambda = \text{photopic}$$
$$V_\lambda' = \text{scotopic}$$

Despite the gross difference in absolute sensitivity between rods and cones, the two functions are plotted on the same graph by performing a simple arithmetic adjustment to put them on a relative basis.

$$\text{relative sensitivity} = \frac{1}{T_\lambda} \times T_{\max} \times 100\%$$

The cone sensitivity curve (Figure 1-10) must be raised a vertical distance of approximately 1.6 log units to make it comparable to the rod sensitivity curve on a relative basis. The relative sensitivity curves are used for photometric problems to represent the response of a standard human observer.

The Purkinje effect is a shift in the maximum sensitivity of the eye from photopic to scotopic vision. The relative spectral sensitivity curve for cones (photopic curve) peaks at 555 nm. During rod or scotopic vision, the relative spectral sensitivity curve shifts 45 nm toward the 400-nm (blue) end of the spectrum so that the peak occurs at 510 nm. This shift results in an increase in sensitivity to shorter wavelengths (400 nm) and a decrease in sensitivity to longer wavelengths (700 nm) for the rod system. Even though objects will be colorless

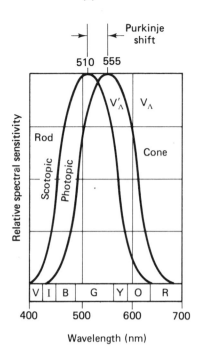

Figure 1-11.   Relative Sensitivity

under rod vision, blue objects of equal reflectance to red objects will appear brighter or more intense during scotopic (rod) vision owing to this shift in sensitivity.

COLOR VISION

Normal color vision is referred to as **trichromatic** vision, meaning three colors. Color vision is due to a visual system that consists of three different classes of cones. Each cone contains a different pigment, and each pigment has a different spectral response curve. The three spectral response curves for each pigment overlap across the visible spectrum. Each spectral response curve is shifted with respect to the other along the wavelength continuum so that each has a different maximum sensitivity. The peak or maximum sensitivities occur at 570, 535, and 445 nm.

However, as can be seen in Figure 1-12, all cones have some sensitivity at all wavelengths. The color receptors are commonly referred to as the red, green, and blue cones. The designations refer to the wavelengths associated with each cone system. According to this theory, normal color vision is a function of the three kinds of pigments present in the three classes of cones. The normal human visual system consists of four classes of receptors, three associated with the cones and

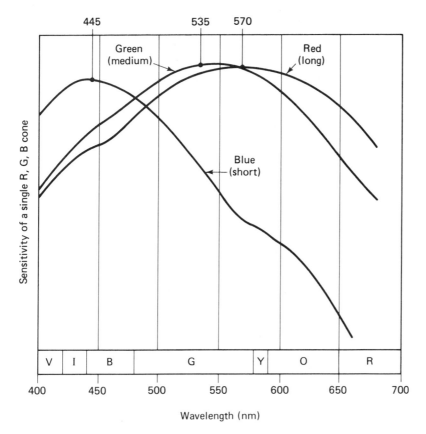

Figure 1-12. Three-Cone Spectral Response (Original drawing by Louis O. Harvey, based on data by Vos and Walraven.)

the fourth associated with the rod pigment (rhodopsin). Each of the four receptors has its own spectral sensitivity curve, which means that the human visual system is actually tetrachromatic rather than trichromatic.

VISUAL PHENOMENA

Afterimage

Afterimage is the continuation of the visual sensation after the stimulus has ceased. It is a continuation of the photochemical process resulting in nerve impulses, which causes object perception for a short period of time. Afterimage effect is proportional to the luminance of the object. Afterimage can be a negative or complementary process. The original color of the object is replaced by its complementary color. Afterimage makes a series of still pictures appear to move, a phenomenon from which motion pictures are derived.

Time Sensitivity

The time sensitivity of the visual system is phenomenal. The human eye can detect a pulsation of light that acts for only $\frac{1}{1000}$ second (s). It can detect a second pulsation if the time interval between the two pulses is $\frac{1}{10}$ s. This phenomena applies only for cone vision, since the cones are much more time sensitive than the rods.

Stimulus Sensitivity

Rods are a thousand times more sensitive to intensity than the cones are at low levels of illumination. Under scotopic vision, it is said that the rods can detect the flicker of a candle 14 miles away.

BINOCULAR VISION

The binocular vision field is the total visual field seen by both eyes, minus the areas blocked by the nose, eyebrows, and the cheeks. The binocular visual field occurs out in the object space. Object space is the real world in front of the observer, as contrasted to the visual space physically inside the brain.

Under normal vision, both eyes focus on an object, and the resulting two images are fused and interpreted by the brain as one. The visual field of both eyes includes more area than the field of one eye alone. The horizontal field subtends an angle of 180°; the average vertical field subtends an angle of 130°(Figure 1-14).

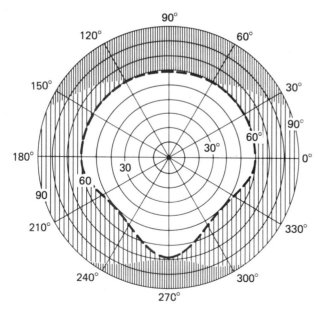

Figure 1-13. Binocular Field of View (Courtesy of Holophane, Inc., H. L. Logan, 1939.)

16

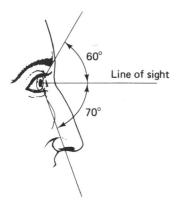

Figure 1-14. Vertical Limits of the Field of View

The central field of distinct vision or foveal vision (Figure 1-6) subtends an angle of 1.4° in object space. The field surrounding the central visual field (1.4 to 20°) has a level of sensitivity that is about 1% of the sensitivity of foveal vision. Beyond the 20° limit, vision is very indistinct and blurred. However, this outer region is very sensitive to changes in brightness and motion.

VISUAL ACUITY

Visual acuity is defined as a measure of the ability of the eye to distinguish detail. Four factors affect visual acuity: (1) size, (2) luminance, (3) contrast, and (4) time.

Size

Size is the most generally recognized and accepted factor in seeing. Visual acuity is dependent upon the size of an object, which affects the size of the image on the retina. The important aspect of size is not the physical size of the object, but the visual angle that the object subtends at the eye. When an object is brought closer to the eye, we are actually increasing the visual angle and making the object clearer (see Figure 1-15).

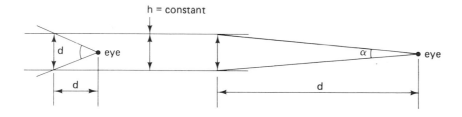

Figure 1-15. Physical Size versus Visual Size

Visual Angle

The visual axis (BB'; see Figure 1-16) meets the retina at the fovea centralis. To determine the size of the retinal image produced by any external object, we use

$$\tan \alpha = \frac{\overline{AB}}{R}$$

For practical purposes, R is measured from the front surface of the cornea. For small visual angles,

$$\tan \alpha = \alpha \text{ (radians)}$$

then

$$\alpha = \frac{\overline{AB}}{R} \text{ (radians)}$$

or

$$\alpha = \frac{180\ \overline{AB}}{\pi R} = \frac{57.30\overline{AB}}{R} \text{ (degrees)}$$

then

$$\alpha = \frac{180}{\pi} \times 60\,\frac{\text{min}}{\text{deg}} \times \frac{\overline{AB}}{R}$$

$$\alpha = \frac{3438\ \overline{AB}}{R} \text{ (minutes)}$$

If $AB = h$, height of object

and $R = D$, distance from the object to the front surface of the cornea

Then, visual angle $\quad \alpha = \dfrac{h}{D} \times 3438$ (minutes)

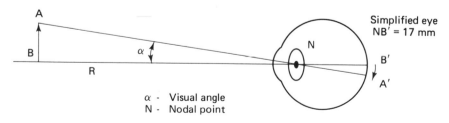

α - Visual angle
N - Nodal point

Figure 1-16. Visual Angle

Actually, D should be measured to the nodal point, which would add approximately 7 mm to the distance. However, in most cases this distance is ignored.

Visual Acuity

The **visual acuity number** is defined as the reciprocal of the minimum visual angle, in minutes of arc, required to detect some critical detail. The larger the angle α required to see an object, the poorer the visual acuity, and hence the smaller the visual acuity number. Increasing the visual angle is achieved by bringing objects closer to the subject.

$$\text{visual acuity number } \nu = \frac{1}{\alpha} = \frac{0.00029D}{h}$$

Snellen found that the threshold size of detail, for most subjects with so-called **normal** vision, was about 1 min of arc for black objects on white background.

The **Landolt ring** or C (see Figure 1-17) has been used as a standard test object for establishing visual acuity. A 1-min Landolt ring will have a critical detail of 1 min with a stroke of 1 min and an overall dimension of 5 min.

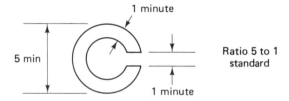

Figure 1-17. Landolt Ring or *C*

Snellen Fraction

The Snellen fraction is

$$\frac{d}{D} = \frac{\text{actual distance}}{\text{normal distance}}$$

where d = distance in feet at which a given line of letters is barely recognizable by any subject

D = distance in feet at which the same line is barely recognizable by a person with normal vision

The standard Snellen test chart is based on a 20-ft viewing distance; therefore, the distance d from which the test chart is actually viewed is set equal to 20 ft. A Snellen fraction of 20/40 means that the subject can barely read a letter at 20 ft that a person with normal vision can see at 40 ft.

Image Height on the Retina

A 1-min object will produce an image height of 0.0049 mm on the retina.

$$h_i = \frac{nh}{D}$$

h_i = height of retinal image
n = nodal point in millimeters = 17 mm
h = height of object in millimeters
D = distance from object to nodal point in millimeters

where

If $\alpha = \frac{h}{D} \times 3438$

then

$$h = \frac{\alpha D}{3438}$$

Therefore,

$$h_i = \frac{n}{D} \times \frac{\alpha D}{3438} = \frac{n\alpha}{3438}$$

For a 1-min test object, $\alpha = 1$ min; then

$$h_i = \frac{17 \times 1}{3438} = 0.0049 \text{ mm} = 4900 \text{ nm}$$

Luminance

Brightness is a subjective evaluation or interpretation of luminance. **Luminance** involves objective evaluation. Luminance is dependent on the amount of light striking a surface and the amount of light being reflected back to the eye. Surfaces with low reflective values require more light than those with high reflective qualities if the visibility is to be equal. For diffuse surfaces,

$$L = \rho \times E$$

where L = luminance
ρ = reflectance
E = illumination

The visual system sees luminance (the energy reflected from surfaces), not illumination. Therefore, calculational procedures based upon the quantity of light (illumination) reaching a surface will give little indication of how well a task can be seen. Seeing is a function of the quality of the light leaving a surface (luminance). The eye receives bundles of light from objects of different

20

luminance in space and must accommodate to these luminance differences by an averaging process.

Comfort and visibility are dependent upon the luminance patterns within the visual field. Comfort is dependent on the variation in luminances in the visual field (luminance ratios). Visibility is affected by the adaptation state, quality of luminance patterns, and clutter within the visual field. Clutter can create an overload of the visual system due to excessive luminance ratios.

Contrast

Contrast is the basic seeing mechanism in vision. **Contrast threshold** is a measure of the ability of an observer to distinguish a minimum difference in luminance between two areas a given percentage of the time. The contrast threshold is expressed as a fraction:

$$C = \left| \frac{L_o - L_b}{L_b} \right| = \left| \frac{\Delta L}{L_b} \right|$$

where L_o = luminance of test object
  L_b = luminance of the background
  Δ_L = is either an increment or decrement superimposed on the background

Figure 1-18 defines contrast terms.

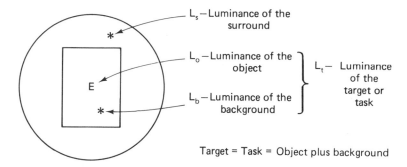

Figure 1-18.   Definition of Contrast Terms

Surface Contrast

Surface contrast is associated with the contrast between an object (print) and its background (paper). The ability to see surface contrast is affected by physical contrast and the surrounding luminance, L_s.

Outline Contrast

If the surrounding luminance L_s is much greater than the target luminance L_t, the task contrast will be lost and the target will go into silhouette. If surrounding luminance L_s is much less than target luminance L_t, the task contrast may not be affected by excessive luminance but may create discomfort, and hence a reduction in visibility.

To see surface detail, that is, to have good surface discrimination, the luminance ratios in the visual field must be controlled. The maximum visual efficiency occurs when the surrounding luminance L_s, has a ratio to target luminance L_t that ranges from $\frac{1}{10}$ to 1.0.

You will notice in Figure 1-19 that visual efficiency falls off rapidly as surrounding luminance exceeds target luminance. Under these conditions the target is in silhouette against the surrounding luminance.

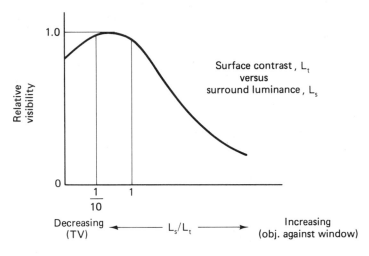

Figure 1-19.   Effect of Surround on Surface Contrast

Surface Contrast, Task Contrast, or Contrast

There are two ways of specifying contrast. The first is the classical approach developed by Weston.

$$C = \frac{L_g - L_l}{L_g}$$

where L_g = the greater luminance
L_l = the lesser luminance

The Weston formula results in contrast limits of C from 0 to 1.

Blackwell found that visibility is not the same if the relationship between object and background is reversed in the Weston formula. If

$$L_o = 5 \ \text{fL} \ \ L_b = 80 \ \text{fL}, \ \ C = 0.94$$

if

$$L_o = 80 \ \text{fL} \ \ L_b = 5 \ \text{fL}, \ \ C = 0.94$$

That is, the contrast values are the same but the visibility is not the same. Therefore, the formula for specifying contrast (developed by Blackwell) is the standard method for calculating contrast.

$$C = \left| \frac{L_o - L_b}{L_b} \right|$$

$$C = \left| \frac{E\rho_o - E\rho_b}{E\rho_b} \right|$$

If E is a constant, then for a Lambertian surface

$$C = \left| \frac{\rho_o - \rho_b}{\rho_b} \right|$$

The Blackwell formula results in the following limits of contrast. If

$L_b > L_o$, C goes from 0 to +1
(L_o, black print; L_b, white paper)
If $L_o > L_b$, C goes from 0 to +∞
(L_o, white print; L_b, black paper)

The highest visibility occurs when the object is brighter than the background ($L_o > L_b$, white print on black paper). The Blackwell method relates directly to visibility. With this formulation, a task with a contrast of 2 will be twice as visible as a task with a contrast of 1. For example, with white print (80 percent) on black paper (5 percent), Blackwell:

$$C = \left| \frac{\rho_o - \rho_b}{\rho_b} \right| = \left| \frac{0.80 - 0.05}{0.05} \right| = \left| \frac{0.75}{0.05} \right| = 15.0$$

Weston:

$$C = \frac{\rho_g - \rho_l}{\rho_g} = \frac{0.80 - 0.05}{0.80} = \frac{0.75}{0.80} = 0.94$$

Contrast Sensitivity

Contrast sensitivity is the reciprocal of contrast. It is the level of sensitivity of the visual system to contrast. If the total quantity of light reaching the retina increases, the sensitivity to surface contrast will increase (see Figure 1-20).

If we superimpose the contrast and contrast sensitivity curves, we can see their reciprocal relationship. As the level of contrast sensitivity increases, the visual system needs less contrast for a certain level of visibility (see Figure 1-21). Points along the contrast curve represent some predetermined level of visibility or equal seeability.

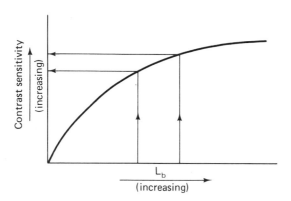

Figure 1-20.   Contrast Sensitivity versus L_b over Normal Indoor Ranges

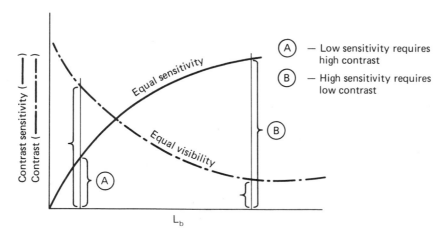

Figure 1-21.   Contrast Sensitivity versus Contrast

Over the normal range of background luminances, contrast sensitivity increases with an increase in background luminance (see Figure 1-22). At very high background luminance, the eye becomes less tolerant of luminance differences between an object (L_o) and its background (L_b), owing to an increase in eye sensitivity, and contrast sensitivity may drop off.

Time

Time is the fourth factor that affects visual acuity. Seeing is not instantaneous. A time lag exists in the electrochemical processing of the retinal signal that reaches the brain. As the level of background luminance increases, the time required to interpret details will decrease. Just as the camera requires a longer ex-

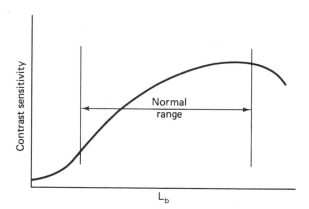

Figure 1-22. Contrast Sensitivity Extended Range

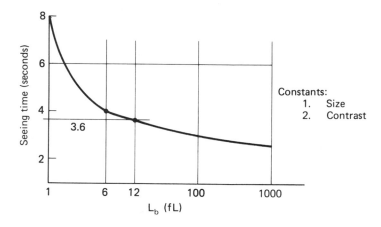

Figure 1-23. Time versus L_b

posure time in dim light than in bright light, so does the eye. The eye can distinguish and discriminate details at low luminance levels if given enough time.

Figure 1-23 shows that, as luminance increases from 1 to 6 fL, seeing time is cut in half; an additional increase from 6 to 12 fL will result in an additional decrease in seeing time of 5 percent.

Conclusions

Visual acuity is affected by four factors: (1) size, (2) luminance, (3) contrast, and (4) time. The lighting engineer has little or no control over size or time. If these two variables are fixed, the ability to see surface detail (visual acuity) is dependent on task contrast and background luminance.

For Constant Contrast.   As the level of background luminance, L_b, increases, the sensitivity to contract increases, which results in an increase in visual acuity (see Figure 1-24).

For Constant Luminance.   At a constant level of background luminance, L_b, the task contrast must be increased to increase the level of visual acuity.

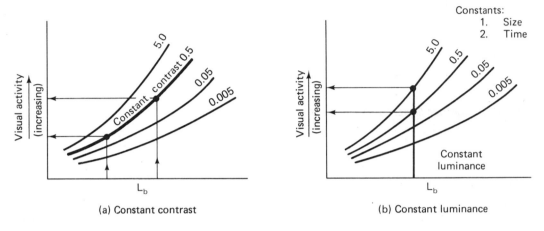

(a) Constant contrast   (b) Constant luminance

Figure 1-24.   Visual Acuity versus L_b

2 color

The subject of color is important to the lighting engineer because of the impact of light on color perception. Color is not a property of objects, but a psychological response to the different wavelengths of radiant energy incident on the retina. The spectral (color) distribution characteristics of objects will produce a specific color appearance for the object. If the spectral characteristics of the light source or object change, the color appearance will change. These interactions must be understood by the lighting engineer if the color appearance desired by the architect or interior designer is to be achieved. The lighting engineer must also make the interaction known to the architect or interior designer. If color selection is not made with the light source in mind, the brilliant, rich, pure colors chosen by the designer may very well appear dull, drab, and washed out. The time, effort, and cost that went into color selection and decor will all be wasted.

Color is also important because of its possible effects on visual acuity. Color is considered by some to be the fifth factor that affects visual acuity. Very little research has been done in the United States on the subject of color contrast. It is obvious that color contrast is important and under certain conditions will affect visual acuity. If two objects have exactly the same texture and reflectance properties but differ in color, they can be distinguished from one another because of color contrast. However, if one is composed of "pure" (single-wavelength spectral distribution) red pigment and the other is "pure" green pigment, and they are both illuminated with monochromatic (single wavelength) yellow light, the two objects will appear as a single object since the contrast between the two is zero.

COLOR THEORY

In the seventeenth century, Newton found that "white" light consisted of many different-colored light rays. When he passed sunlight through a prism of glass, the light ray was dispersed into a rainbow of colors he called the **spectrum**. By passing the light spectrum through a second prism, the energy emerged from the second prism as a single ray of "white" light. Newton concluded from his experiments that the seven colors in the spectrum were the fundamental or primary colors.

If a single wavelength of light of sufficient quantity, such as 650 nm, strikes the retina of the eye, the sensation of distinct color is produced and the stimulus is described as red. Color has been shown to be a sensation or a matter of vision. The sensation of color (color vision) is produced by the action of radiant energy of specific wavelengths acting on the retina of a normal eye. The radiant energy leaves the light source in the form of waves traveling at the speed of light. Variations in the wavelength of the radiant energy will produce different sensations that correspond to different hues or colors.

Newton's experiments showed that the nature of light (its spectral distribution characteristics) influences the perception of color. That is, objects will look different under daylight, incandescent, or fluorescent light. Object color is dependent on the nature of the light cast on the object, the reflectance of the object, and the response characteristics of the subject's eye. Object color is due to the phenomenon called **selective absorption** (see Figure 2-1). Selective absorption is the result of object pigment decomposing the light particles illuminating the object, absorbing some of the rays, and reflecting or scattering others. The properties of the material or its pigmentation determines the color appearance by a sub-

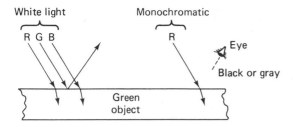

Figure 2-1. Selective Absorption

tractive or absorptive process. This process allows only certain wavelengths to be reflected, which gives an object its color appearance. A gray object is produced by absorbing a certain percentage of rays and reflecting the remaining portion without disturbing the relative proportion of energy in the source.

Subtractive Color Mixing Theory

Brewster, experimenting with pigments and dyes, found that three colors existed that could be mixed to form the seven colors Newton had identified in the spectrum. The three colors, which Brewster referred to as the **primaries,** were magenta, yellow, and cyan (red, yellow, and blue). Brewster's theory was in accordance with other research conducted by scientists and colorists.

The primary colors of pigments (Figure 2-2) are called the **subtractive primaries,** which are magenta (red), yellow, cyan (blue). By mixing pairs of subtractive primaries, the subtractive **secondaries** are formed: red, blue, green. *Complementary* colors are across from each other in Figure 2-2. For example, blue is referred to as the complement of yellow, or vice versa. A mixture of all three primaries results in black or the absence of color.

Subtractive color mixing theory (pigment or dye primaries)

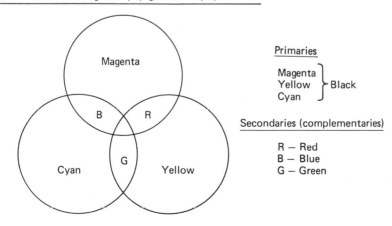

Figure 2-2. Subtractive Primaries

Additive Color Mixing Theory

In the nineteenth century, Thomas Young formulated the theory that white light is composed of three primary colors (red, green, blue), and that these colors mix to produce the other seven colors of the spectrum. The Young theory was not generally accepted until Helmholtz and Maxwell confirmed it.

Helmholtz, expanding on Young's work, theorized that three groups of nerve fibers or photochemical substances are present in the eye. Each group of nerve fibers is sensitive to one of the three light primaries; that is, one group is sensitive to red, one group to green, and the third group to blue. Helmholtz and Young felt that intermediate hues were derived from action on at least two of the three nerve-fiber groups. Therefore, light that simultaneously affects the nerve fibers sensitive to red and green produces the sensation of yellow. The Helmholtz–Young theory was contrary to Brewster's theory, which stated that yellow was a primary color. According to the Helmholtz–Young theory, white light would be produced by equal stimulation of all three nerve groups.

The Helmholtz–Young theory still prevails for spectral color mixing. This mixing of colored lights (Figure 2-3) is called the **additive** color process. When two groups of wavelengths are added, the result is a third color. By mixing the **additive primaries**, the secondary colors are produced. Red and green produce the secondary color yellow; red and blue produce magenta (red-purple). Green and blue produce the secondary cyan (blue-turquoise); red plus blue plus green will produce white light.

Complementary colors are secondary and across from the primaries on the color circles. Yellow (a secondary produced by mixing red and green) is the complement of blue (a primary). The light secondaries are almost identical with the pigment primaries. White light is achieved by mixing the three primaries of light,

Additive color mixing theory (light primaries)

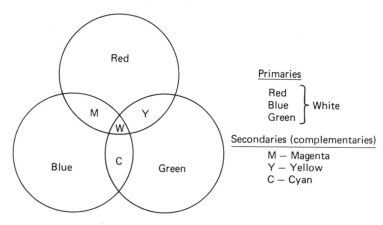

Primaries

Red ⎫
Blue ⎬ White
Green ⎭

Secondaries (complementaries)

M — Magenta
Y — Yellow
C — Cyan

Figure 2-3. Additive Primaries

and black represents the absence of light, whereas black pigment is a subtraction of the three primaries of light.

Psychological Effects of Color

Colors have long been known to have psychological and emotional effects on man. Some of the psychological and emotional terms that have been used to define these effects and the color associated with them are as follows:

1. Warm, advancing, and stimulating colors: red, orange-red, orange, yellow.
2. Cool, receding, relaxing colors: blue-green, violet.
3. Neutral, tranquil colors: yellow-green, green. Neutral is equidistant between the ends of the spectrum.

Shades, which are achieved by the addition of black, are considered warmer than pure colors. **Tints,** which are achieved by the addition of white, are considered cooler than pure colors. Gray may be neutral, cool, or warm depending on whether it is tinted or shaded.

Red is associated with health, power, anger, fire, and heat; red-purple with royalty; reddish-brown with harvest and fall. Yellow is associated with the sun (warmth) and is said to be one of the most cheerful colors. Gold is associated with richness and splendor. Green is associated with nature, life, and the out-of-doors. Blue is associated with constancy, fidelity, and is cool. White is associated with light and purity; black is associated with darkness, death, and gloom.

These associations are important facts that should be kept in mind where mood and atmosphere are to be created. The existence of these strong color associations can be proved by conducting a simple word/color association experiment. A subject is asked to respond with the first color name that the subject associates with the sample words shown in Table 2-1.

TABLE 2-1

Color Association Sample Test

Fire (red)[a]	Death (black)
Richness (gold)	Power (red)
Fall (reddish-brown, orange)	Splendor (gold)
Cool (blue)	Purity (white)
Cheerful (yellow)	Nature (green)
Fidelity (blue)	Warm (red)
Anger (red)	Royalty (red-purple, violet)

[a]Most common answer is in parentheses.

Color Contrast

Chromatic areas are affected by the adjacent or surrounding area colors. A color patch will appear brighter or less gray if the background color is relatively dark. The same patch will look dimmer or more gray if the background color is relatively light.

COLOR CLASSIFICATION

A number of color classification systems have been proposed over the years. Three of these classification systems will be presented here: (1) Munsell system, (2) Ostwald system, and (3) CIE chromaticity system.

The Munsell and Ostwald systems are used to denote pigment colors for a standard lighting condition. Either of these two systems can be used when ordering paints. A specifier placing an order for a light orange paint could get any one of a thousand tints of orange. However, by using one of the color notation systems, such as 5YR7/8 (Munsell), the specifier would be assured of getting the desired color.

The CIE chromaticity system is used primarily in the field of color research, manufacturing, processing, and marketing. The CIE chromaticity system has the advantage of combining the effects of object color, source color, and the visual system to determine the actual perceived color under the given conditions. True object color can be predetermined for a given light source color.

Munsell Color System

The Munsell color system was conceived in 1898 by A. H. Munsell, a lecturer in color composition and art at the Normal Art School in Boston. Munsell's basic reason for developing the system was to aid in teaching his color composition class. The system would enable him to describe the colors on his sketches in definite terms. Munsell began by arranging colors on a circle, then on a sphere. From this beginning, the Munsell color system was developed.

The Munsell color system is broken into three parts: hue, value, and chroma. **Hue** is the name of a color: red, blue, and so on. **Value** is a measure of the lightness or darkness of a color. The higher the value number, the lighter the color and the higher the reflectance. Lower-value numbers indicate darker colors and lower reflectance. **Chroma** is a measure of the saturation (or purity) of a color. The spectral colors have the maximum chroma, or are 100 percent pure. By adding gray, a color becomes less saturated until it finally has no color, or is achromatic (gray).

A segment from the system (Figure 2-4) would show all three parts and should be perceived under ordinary viewing conditions, that is, by daylight illumination. The colors progress from very dark at the bottom of each chart to very light at the top. They vary by steps that are intended to be visually equal. Horizontally, they progress from achromatic colors (black, gray, or white) at the left side of each chart to saturated colors on the right side by steps that are also intended to be visually equal.

Constant-Hue Chart (Figure 2-4)

The locus of colors producible by mixtures of black, white, and pigment would be indicated by curved lines, such as the dotted line of Figure 2-4. The system is built up at each hue by adding progressive columns of colors to the nine

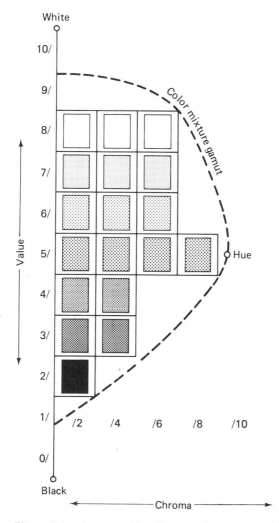

Figure 2-4. Constant Hue Segment (Courtesy of Munsell Color)

neutrals of the same value. Each column of colors departs from the equivalent neutral by two additional steps of chroma. Each row of colors of constant value is extended until the next interval of two chroma steps would extend beyond the colorant-mixture gamut.

Constant-Value Chart (Figure 2-5)

The 100-point Munsell hue scale and notation are shown around the outer circle of Figure 2-5. Colors of constant hue are shown by the radial lines intersecting at the point that represents the neutral or gray scale (chroma /0). The hue scale is built up of 10 segments of 10 hues each, such as the segment from 1R to 10R. The fifth hue of each of these segments is noted by the letters alone. Thus

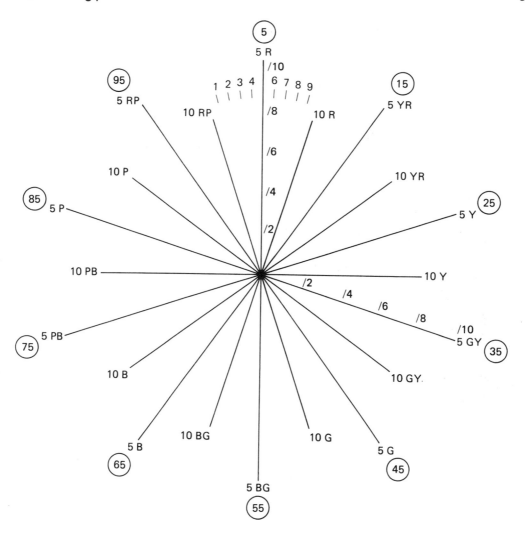

Figure 2-5.   Constant Value Segment

5R is simply R. The chroma scale up to chroma /10 is shown along any radial line in the figure, and the colors of constant chroma are shown by concentric circles.

Munsell Solid (Figure 2-6, 2-7)

If the constant-hue and constant-value charts were shown in three-dimensional form, the resultant figure would be a distorted sphere (Figure 2-6).

Notation System

Each individual color chip in the Munsell system is denoted by three symbols, for example, 5G 7/4. Here 5G indicates hue-5 green, 7 indicates value, and 4

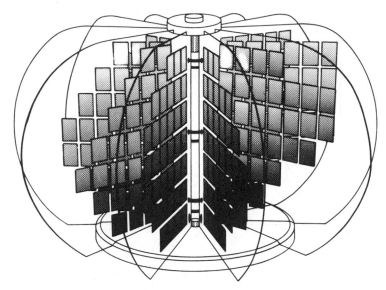

Munsell color tree illustrating color solid

Figure 2-6. Munsell Color Solid (Courtesy of Munsell Color)

indicates chroma. The color 5G 7/4 could also be denoted according to the 100-hue divisions, in which case the color would be specified as 45-7/4 (Figure 2-5). Thus, by using the Munsell color notation system any color can be accurately defined.

Conclusions

The principal use of the Munsell color system is for color specification, since the system lends itself so well to this function. However, it is also used for color education.

The continued availability of the Munsell samples was due to the Munsell Color Foundation until 1970. In 1970, Munsell Color became a part of the Macbeth Division of Kollmorgen Corporation, 2441 North Calvert St., Baltimore, Maryland, 21218. A catalog listing available materials and equipment, such as "student sets," can be obtained from Munsell Color.

Ostwald Color System

The Ostwald color theory was developed in 1914 by Wilhelm Ostwald. He selected four evenly spaced primaries to develop a color wheel. Ostwald's primaries were lemon-yellow, scarlet, ultramarine, and turquoise green.

Ostwald divided his original color wheel into 100 equally spaced hues. He found this to be impractical, since it was difficult to distinguish one color step from another. He finally decided to divide the color wheel into 4 primary colors and 5 intermediates, which formed a color wheel of 24 distinct hues.

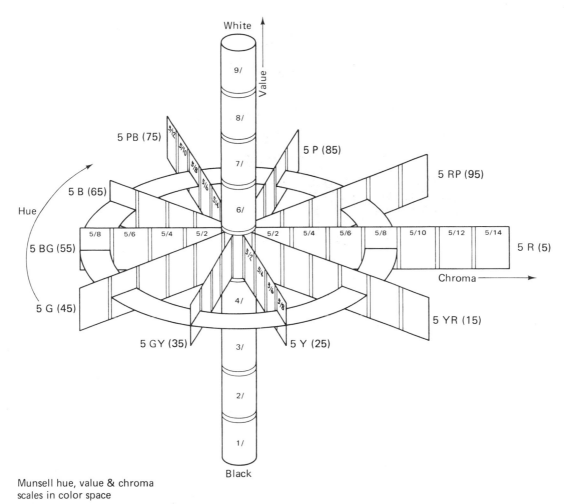

Munsell hue, value & chroma
scales in color space

Figure 2-7.   Munsell Color Solid (Courtesy of Munsell Color)

Achromatic Scale

Gustav Fechner, a German scientist, developed experiments that indicated that between black and white our perception is stimulated by a logarithmic progression of gray steps. Based on Fechner's studies, Ostwald organized a gray scale based on a uniformly varying proportion of white and black. His first study consisted of a scale with 25 stops, which he lettered with the alphabet, omitting the letter J. He concluded that a more practical scale needed only 8 gray stops. Each stop is given a letter designation: A, C, E, G, I, L, N, P. These 8 stops represent the **achromatic** scale (or gray scale) in the Ostwald System.

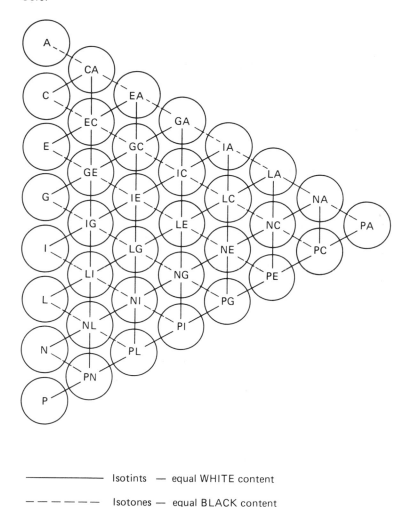

——————————— Isotints — equal WHITE content

— — — — — — Isotones — equal BLACK content

Figure 2-8. Ostwald Constant Hue Segment

Color Solid

With the 8 grays and 24 hues, the resulting solid contains 680 colors. Each hue segment is an equilateral triangle with each color designated by letter combinations (see Figure 2-8). The first letter indicates the percentage of **white** content, the second letter the percentage of **black** content. Therefore, colors with like first letters have equal white content and are called **isotints.** Those with like last letters have equal black content and are called **isotones.** Colors around the color solid with the same pair of letters are called **isovalents.** Isovalents have the same value and chroma but differ in hue.

Color Notation

The three-symbol notation system defines the percentages of white, black, and color that each of the color chips contains. Table 2-2 gives the percentages of white and black associated with each letter value.

TABLE 2-2

Ostwald Color System: Percentage of White/Black

Ostwald value	White content	Black content
A	0.8913	0.1087
C	0.5623	0.4377
E	0.3548	0.6452
G	0.2239	0.7761
I	0.1413	0.8587
L	0.0891	0.9109
N	0.0562	0.9438
P	0.0355	0.9645

Example: Ostwald colors 7LE and 20EA.

7 = scarlet
L = First Letter = % white = 0.089
E = Second Letter = % black = 0.645
% color + (%W + %B) = 0.266

20 = turquoise green
E = % white = 0.355
A = % black = 0.109
% color = [1 − (%W + %B)] = 0.536

Note that the extreme color PA equals only 0.8558 percent [1 − (0.0355 + 0.1087)] color saturation. This indicates that, according to the Ostwald theory, full color free from black and white represents an ideal that cannot be realized by any form of pigment or dye.

Harmonies

An advantage with the Ostwald system is in its application to color harmony. The Ostwald solid is symmetrical and orderly which makes the Ostwald system a more reliable means of choosing color harmonies. The 680 colors are equally spaced and therefore are geometrically and mathematically related, as well as visually related. The mathematical and geometrical harmony obtainable can be compared to the chords possible in music. That is, equal intervals can be chosen or intervals removed from each other by mathematical progression can be used.

CIE Chromaticity System

The International Commission on Illumination (Commission Internationale de l'Eclairage) has chosen as a basis for standardization the response of the three sets of color receptors in the eye. The CIE system uses a tristimulus method of specifying the color of light or a surface in terms of the amounts of the primary colors of light required to match the unknown color in question. The system employs a chromaticity diagram that contains the seven colors of the visible spectrum (Figure 2-9). The spectral colors are located on the locus (edge) of

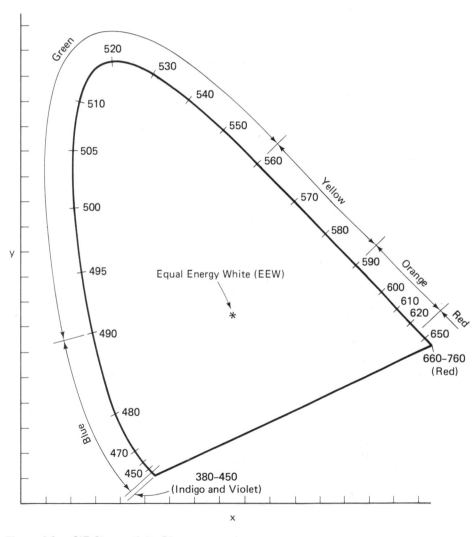

Figure 2-9. CIE Chromaticity Diagram

the chromaticity diagram. A straight line (purple boundary) joins the red and violet ends of the spectral locus to form a closed diagram.

Terms and Definitions

X, Y, and Z are **tristimulus values** that represent the absolute values of the red, green, and blue primaries.

x, y, and z are **chromaticity coordinates or trichromatic coefficients** that represent the fractional values of the three primaries.

$\bar{x}$, $\bar{y}$, and $\bar{z}$ are **distribution functions** that represent the tristimulus values—the red, green, and blue response of the standard eye.

Standard Light Sources

The CIE system is based on three standard light sources:

Source	Description	Chromaticity coordinates
Standard A	Tungsten lamp	$x = 0.4476$ $y = 0.4075$
Standard B	Mean noon sunlight	$x = 0.3485$ $y = 0.3518$
Standard C	Average daylight	$x = 0.3101$ $y = 0.3163$

General

The object color to be specified (unknown color) is subjected to a spectrophotometric measurement to determine its reflectance (or transmittance) properties to each wavelength across the spectrum. The specification of an unknown color is also dependent on the spectral characteristics of the light source that the object is being viewed under.

The distribution functions of the three tristimulus values (x, y, z) have been mathematically transformed so that the tristimulus value y (green) matches the standard luminosity curve (see Chap. 1). The distribution functions are each mathematically combined with the unknown color's spectral reflectance curve and the spectral energy distribution of the light source to define the absolute tristimulus values (X, Y, Z).

The fractional values of red (x) and green (y) of the mixture to match the unknown color can be found by the equations

$$x = \frac{X}{X + Y + Z}, \qquad y = \frac{Y}{X + Y + Z} \quad \text{(chromaticity coordinates)}$$

Since $x + y + z = 1$, only two fractional values are required to define an unknown color. As a result, the chromaticity diagram (Figure 2-10), which is used to plot the chromaticity coordinate, can be a two-dimensional diagram. The fractional value of red (*x* chromaticity coordinate), is plotted along the abscissa, and the fractional value of green (*y* chromaticity coordinate) is plotted along the ordinate.

The chromaticity coordinates (*x, y*) of the unknown color (object location) and the coordinates of the light course (source location) are plotted (Figure 2-10) on the CIE chromaticity diagram. A straight line is drawn **from** the light source location **through** the object location and extended on **to** the spectral locus of the diagram. The point at which the straight line intersects the spectral locus defines the **dominant wavelength.** The dominant wavelength is the dominant color appearance of the object (unknown color) under that given light source. The excitation purity of the object can be found (Figure 2-10) by determining the proportional length of the line (*ao/ab*) from the light source (*a*) to the object (*o*) and from the light source (*a*) to the locus (*b*) of the diagram.

The maximum color saturation, or 100 percent purity, occurs on the spectral locus of the diagram; zero saturation or no color occurs at the point defined as **equal energy white** (EEW). Purity relative to EEW is found (Figure 2-10) by constructing a straight line **from** EEW **through** the object **to** the spectral locus. Purity (EEW) is determined by the proportional length of the line (*co/cd*) from EEW (*c*) to the object (*o*) and from EEW (*c*) to the locus (*d*) of the diagram.

Characteristics of CIE (*x, y, z*) System

1. All chromaticity coefficients are positive.
2. The *y* value doubles as a measure of luminance (reflectance) and chromaticity coordinate. The *x* axis represents an axis of constant zero luminance referred to as the Alychne axis, so that *y* values are luminance values.
3. *y* (distribution function) is equal to the standard luminosity curve.
4. From 650 to 700 on the spectral locus is a straight line, which represents a *z* value of zero.

Calculation Methods

There are three methods of obtaining the CIE specification of an object of unknown color under a given light source: (1) the graphical solution, (2) the weighted-ordinate method, and (3) the selected ordinate method (10 or 30 ordinates). The more irregular the curve, the greater is the number of ordinates required for accuracy.

GRAPHICAL SOLUTION. A graphical solution is a useful procedure for gaining insight into the workings of the CIE chromaticity system. This is also the most exact procedure and is quite tedious if done longhand. It is the method recommended for the computerized analysis of color using the CIE system.

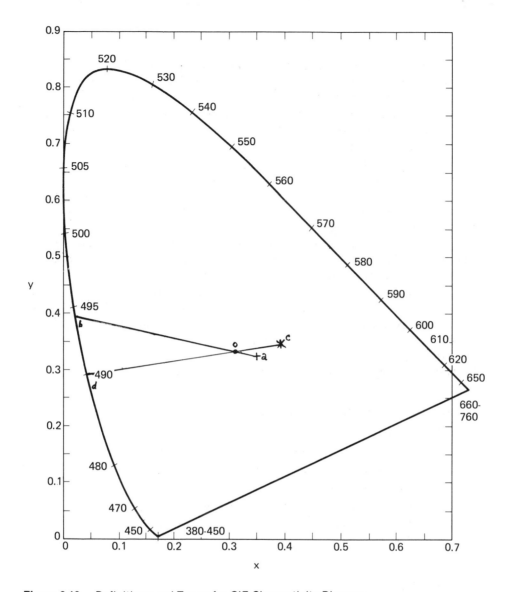

Figure 2-10. Definitions and Terms for CIE Chromaticity Diagram

Example: What is the color appearance (unknown color) of a given object when illuminated by standard source C?

1. Obtain the spectral reflectance distribution data (curve a, Figure 2-11) for the object of interest.

2. The spectral energy distribution (curve b, Figure 2-11) of a given light source (this example uses standard source C) is combined with the spectral reflectance distribu-

42

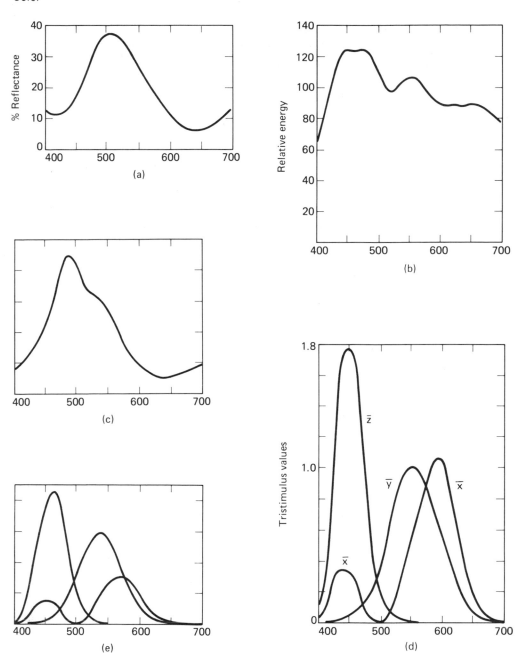

Figure 2-11. Graphical Procedure: Curve (a), Spectral Reflectance Distribution of the Object, Curve (b), Spectral Energy Distribution for Source *C,* Curve (c), Combined Curve, Curve (d), Distribution Functions, and Curve (e), Absolute Values of R, G, B Primaries (Tristimulus Values).

tion of the object (previous). The distribution curves are combined by multiplying the ordinates of the two curves together.

3. Combined distribution curves (curve a × b = curve c, Figure 2-11).

4. The ordinates at each wavelength interval along the distribution functions (curve d, Figure 2-11) are multiplied by the ordinates of the previous combined curve (curve c) at the same wavelength interval.

5. The resulting curves (curve e, Figure 2-11) represent the absolute values X, Y, and Z for the unknown surface color under the specified light source (standard source C).

6. The chromaticity coordinates x and y are obtained by dividing the area under the X or Y by the sum of the areas under the three curves (e). The x and y values can then be plotted on the chromaticity diagram (see Figure 2-10), position o) to determine the CIE specification of the object.

7. The chromaticity diagram can be used to determine the following:
   a. The reflectance of the object under the light source, which is equal to the total area under the Y curve (curve e).
   b. The dominant wavelength or predominant color appearance of the object under the light source.
   c. The excitation purity relative to the light source.
   d. The purity of the object color relative to EEW.
   e. The required mixture of two or three light sources to reproduce the same color appearance as the object.

METHOD OF TEN SELECTED ORDINATES. A procedure has been developed to rapidly determine the CIE specification of an object under the three standard illuminants (source A, B, or C). Preselected ordinates (wavelengths) and weighting functions (column factors) have been selected by the color committee of the CIE. An example problem utilizing the method of ten selected ordinates is given at the end of this chapter.

Source Blending

The matching of a lamp color can be done by mixing the light from two or more other light sources. Any two points joined together with a straight line that passes through the source to be matched (G) can be combined in the suitable proportions to give the desired matching color (see Figure 2-12). Thus the straight line AB through G shows that wavelengths of 460 nm (blue) and 574 nm (yellow) can be mixed in the proportion of $\dfrac{AG}{AB}$ of 574-nm light plus $\dfrac{GB}{AB}$ of 460-nm light give the color G.

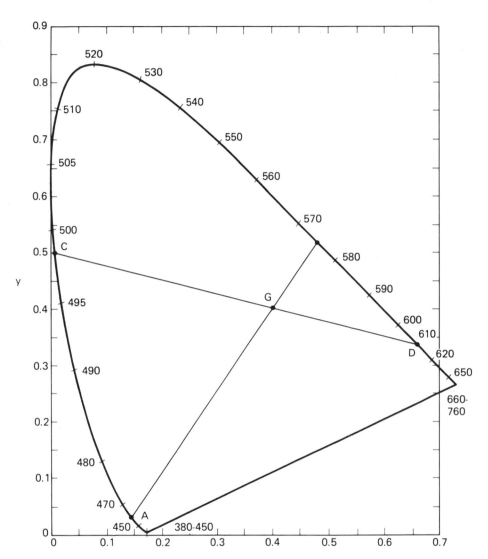

Figure 2-12. Examples of Source Blending

reference

1. Kaufman, John, editor, **IES Lighting Handbook,** 5th ed. Illuminating Engineering Society, New York, 1972.

data section

TABLE 2-3

Ten Selected Ordinates for Sources A, B, and C

Source A: Tungsten, 2854°K, x = 0.448, y = 0.408

Ordinate number	X	Y	Z
1	517	508	425
2	561	530	436
3	576	544	444
4	587	555	451
5	597	566	457
6	605	577	463
7	614	588	469
8	623	600	477
9	635	615	488
10	656	640	508

Source B: Noon sunlight, 4800°K, x = 0.349, y = 0.352

1	442	495	423
2	528	520	433
3	559	534	440
4	573	545	446
5	585	556	452
6	595	566	458
7	605	577	464
8	615	589	471
9	628	605	480
10	649	631	499

Source C: Average daylight, 6740°K, x = 0.310, y = 0.316

1	436	489	422
2	461	515	432
3	544	530	439
4	564	541	444
5	577	552	450
6	589	562	456
7	600	573	462
8	611	585	469
9	624	601	478
10	644	627	495

TABLE 2-3 (cont.)

Column factors	X	Y	Z
Source A	0.110	0.100	0.036
Source B	0.099	0.100	0.085
Source C	0.098	0.100	0.118

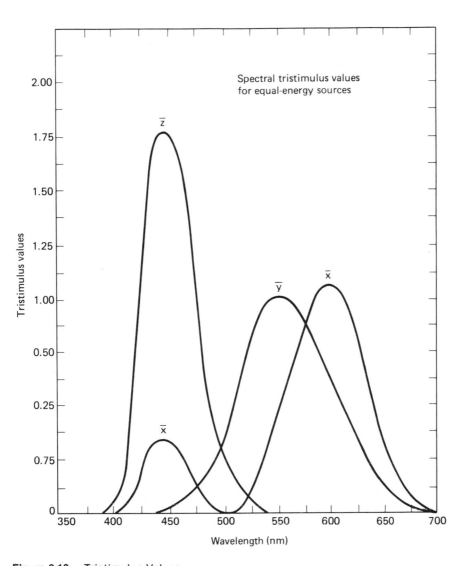

Figure 2-13. Tristimulus Values

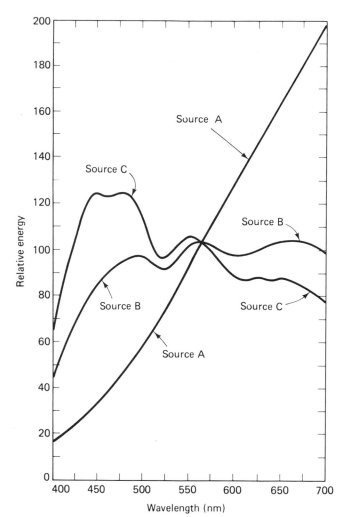

Figure 2-14.   Three Standard Light Sources

Figure 2-15.   Spectral Reflectance for Munsell 5R4

48

example problem

Method of 10 Selected Ordinates

Illuminant ___A___

Spectral reflectance for <u>Munsell 5R4/14</u>

	X		Y		Z	
Ordinate number	λ	λp	λ	λp	λ	λp
1	517	0.040	508	0.045	425	0.050
2	561	0.040	530	0.040	436	0.048
3	576	0.065	544	0.040	444	0.048
4	587	0.165	555	0.040	451	0.045
5	597	0.280	566	0.040	457	0.045
6	605	0.350	577	0.060	463	0.043
7	614	0.550	588	0.165	465	0.043
8	623	0.600	600	0.300	477	0.040
9	635	0.640	615	0.555	488	0.040
10	656	0.675	640	0.650	508	0.040
Sum		3.405		1.935		0.422
× Column factor		0.110		0.100		0.036
	$X =$	0.375	$Y =$	0.194	$Z =$	0.016

$x = \dfrac{X}{X + Y + Z} = \dfrac{0.375}{0.585} = 0.64$ Dominant wavelength = 615 nm

$y = \dfrac{Y}{X + Y + Z} = \dfrac{0.194}{0.585} = 0.33$ Excitation purity $= \dfrac{ao}{ab} = 84.9\%$

Reflectance $= Y = 0.194 = 19.4\%$ Purity EEW $= \dfrac{co}{cd} = 93.6\%$

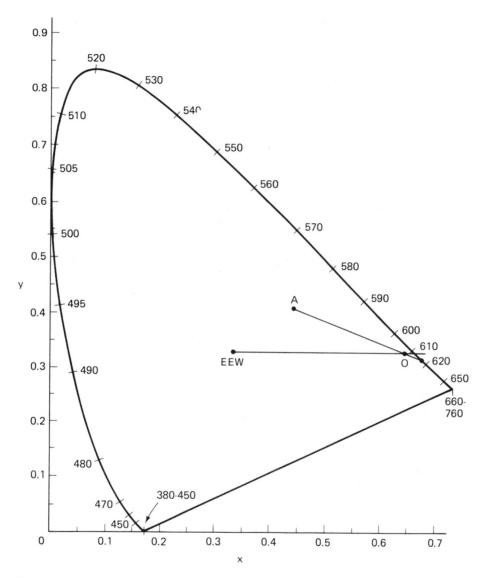

Figure 2-16. Example of CIE Chromaticity Plot

3 photometric units and terminology

LIGHTING TERMS

The photometric units and terminology presented in this chapter are the more commonly occurring basic quantities of value to the illuminating engineer. Conversion factors are at the end of this chapter in the data section.

Luminous energy (Q): visually evaluated radiant energy traveling in the form of electromagnetic waves. The unit is the lumen-second (lm·s).

Solid angle (ω): the ratio of the sphere surface area enclosed to the square of the radius.

$$\omega = \frac{A \text{ sphere}}{R^2} = \frac{dA}{R^2}$$

The unit is the steradian (sr).

Luminous flux (ϕ): the time rate of flow of luminous energy. For example, the light emitted from a 40-watt (W) fluorescent lamp is at a rate of 3250 lumens.

$$\phi = \frac{dQ}{dt}$$

The unit of luminous flux is $\dfrac{(\text{lm} \cdot \text{s})}{\text{s}}$ or the lumen (lm).

The lumen is the flux emitted within a solid angle by a point source having a luminous intensity of 1 candela.

$$\text{lumen} = 1 \frac{\text{candela}}{\text{steradian}} = 1 \frac{\text{cd}}{\text{sr}}$$

Luminous intensity (I): the solid-angular luminous flux density in a given direction. Luminous intensity is commonly (but incorrectly) referred to as candlepower. Luminous intensity indicates the ability of a light source to produce illumination in a given direction.

$$I = \frac{d\phi}{d\omega}$$

The unit is a lumen/steradian, or a candela (cd).

Luminance (L): the luminous flux per unit of projected area (A_θ) per unit solid angle ($d\omega$) leaving a given point in a given direction. The term used to be called *brightness*. Brightness is used to describe a subjective evaluation of a surface, whereas luminance (or photometric brightness) is used to describe the objective or measured characteristics of a surface.

$$L = \frac{d^2\phi}{d\omega dA_\theta}$$

The unit is the lumen/square foot, or the footlambert (fL).

When the unit of luminous flux is the lumen and the projected area is in square feet (solid angle is unitless), the unit of luminance is the footlambert. Luminance can also be defined as the luminous intensity (I) of a surface in a given direction (θ) per unit projected area as viewed from that direction (A_θ).

Since,

$$I = \frac{d\phi}{d\omega}$$

then

$$L = \frac{d^2\phi}{d\omega dA_\theta} = \frac{I_\theta}{dA_\theta}$$

When the luminous intensity is in candelas and the area is in square inches, the unit of luminance is in candelas/square inch. A derivation at the end of this chapter shows how the π factor enters into the following definition:

$$1 \text{ fL} = 1 \frac{\text{lm}}{\text{ft}^2} \times \frac{1}{\pi\frac{\text{lm}}{\text{cd}}} = \frac{1}{\pi}\frac{\text{cd}}{\text{ft}^2} = \frac{1}{144\pi}\frac{\text{cd}}{\text{in.}^2}$$

The amount of footlamberts is equal to $\frac{1}{144\pi}$ times the amount of candelas per square inch.

Conversion formula:

$$\frac{1 \text{ fL}}{(1/144\pi)\,(\text{cd in.}^2} = \boxed{\frac{144\pi \text{ in.}^2 \times \text{fL}}{1 \text{ cd}}} = 1$$

For example,

$$3\frac{\text{cd}}{\text{in.}^2} \times \frac{144\pi \text{ in.}^2 \times \text{fL}}{1 \text{ cd}} = 1357.2 \text{ fL}$$

Lambertian surface (perfect diffuser): a surface that acts like a perfect diffuser. That is, it absorbs all the incidented luminous energy and re-emits all the luminous energy in the form of a tangent sphere, according to Lambert's cosine law (see Figure 3-1). Lambert's cosine law states that the luminous intensity at any angle (I_θ) is equal to the luminous intensity perpendicular (I_N, normal) times the cosine of the angle θ.

$$I_\theta = I_N \cos \theta$$

A Lambertian surface will have a constant luminance irrespective of the direction of view.

$$L = \frac{I_\theta}{dA_\theta}$$

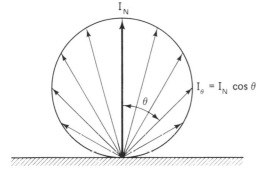

Figure 3-1.   Lambert's Cosine Law (Perfect Diffuser)

Since $dA_\theta = A_N \cos\theta$ = projected area (Figure 3-2), and $I_\theta = I_N \cos\theta$ for a Lambertian surface,

$$L = \frac{I_N \cos\theta}{A_N \cos\theta} = \frac{I_N}{A_N} = \text{constant}$$

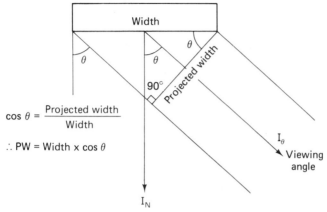

$$\cos\theta = \frac{\text{Projected width}}{\text{Width}}$$

$$\therefore \text{PW} = \text{Width} \times \cos\theta$$

Projected area, A_θ = length x projected width
A_θ = length x width x $\cos\theta$
$A_\theta = A_N \cos\theta$

Figure 3-2. Projected Area

Luminous exitance (M): the luminous flux **emitted** by a very small surface divided by the area of that surface element. Luminous exitance is also known as the density of luminous flux emitted.

$$M = \frac{d\phi}{dA}$$

The unit is the $\dfrac{\text{lumen}}{\text{square foot}}$.

Illumination (E): the **incidented** luminous flux on a small surface per unit area of the surface (Figure 3-3).

$$E = \frac{d\phi}{dA}$$

The unit is the $\dfrac{\text{lumen}}{\text{square foot}}$ or the footcandle (fc).

The **fundamental law of illumination,** known as the **inverse square law** (Figure 3-4) can be derived. Since

$$I = \frac{d\phi}{d\omega}$$

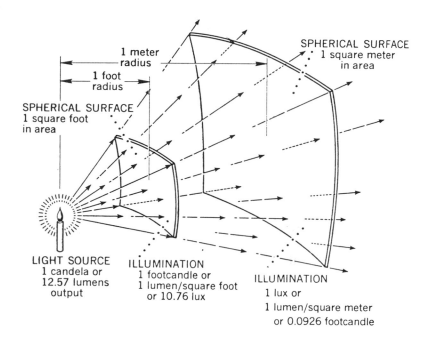

1 meter radius

1 foot radius

SPHERICAL SURFACE
1 square foot
in area

SPHERICAL SURFACE
1 square meter
in area

LIGHT SOURCE
1 candela or
12.57 lumens
output

ILLUMINATION
1 footcandle or
1 lumen/square foot
or 10.76 lux

ILLUMINATION
1 lux or
1 lumen/square meter
or 0.0926 footcandle

Figure 3-3. Illumination (1 ms/ft²; footcandles)

then

$$d\phi = Id\omega$$

and

$$d\omega = \frac{dA}{R^2}$$

then

$$dA = R^2 d\omega$$

therefore,

$$E = \frac{Id\omega}{R^2 d\omega} = \frac{I}{R^2} = \frac{I}{d^2}$$

(since R equals the distance, d, from the light source to the surface area).

Reflectance (ρ): the ability of a surface to reradiate energy. This terminology applies only to a Lambertian surface or perfect diffuser.

$$\rho = \frac{\text{total reflected light (fL)}}{\text{total incidented light (fc)}} \times 100$$

Reflectance is expressed as a percentage.

55

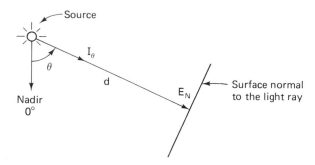

Figure 3-4. Inverse Square Law

Bidirectional reflectance distribution function (BRDF): a term that applies to the reflectance characteristics of any surface that is non-Lambertian and therefore cannot be described by a single value. The solid (called an **indicatrix**) formed by the various reflectance vectors is a function of three variables, h, v_o, and v_s.

h = horizontal angle between the fixed component and the moving component

v_o = vertical angle from a perpendicular to a point on the surface, P, to the line from the observer, 0, (detector, photometer, photocell), to the point on the surface

v_s = vertical angle from a perpendicular to a point on the surface, P, to the line from the source, s, to the point on the surface

To measure the BRDF's the v_o angle is fixed for a given indicatrix and the source is moved about in the hemisphere above the point on the surface (see Figure 3-5).

Reflected luminous intensity indicatrix: reflected luminous intensity for non-Lambertian surfaces is a vector function of three variables (h, v_o, v_s), as defined above. For the RLI indicatrix, the v_s angle is fixed for a given indicatrix and the detector moves about in the hemisphere above the point on the surface (see Figure 3-6).

Transmittance (τ): a measure of the amount of light transmitted through a material expressed in a percentage. This term applies only to a Lambertian transmitter.

$$\tau = \frac{\text{total transmitted light (fL)}}{\text{total incidented light (fc)}} \times 100$$

Transmittance is expressed as a percentage.

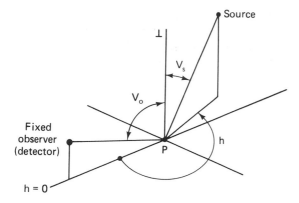

Figure 3-5. Measurement of Bidirectional Reflectance Distribution Function

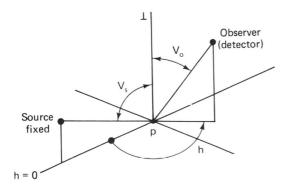

Figure 3-6. Measurement of Reflected Luminous Intensity

Efficacy (ξ): luminous efficacy is a ratio of the total luminous flux emitted by a source to the total power input to the source.

$$\xi = \frac{\text{total luminous flux emitted}}{\text{total input power}}$$

The unit is the $\dfrac{\text{lumen}}{\text{watt}}$ (lm/W).

Spectral luminous efficiency: the ratio of radiant flux at wavelength λ_m to that at any other wavelength such that both wavelengths produce equally intense visual sensations. λ_m is chosen so that the maximum value of this ratio is 1. The spectral luminous efficiency of all wavelengths defines a function that is the response of the human visual system to radiant energy. Symbols are $V(\lambda)$ for photopic vision and $V'(\lambda)$ for scotopic vision. The function of $V(\lambda)$ is commonly known as the **standard observer curve.**

Unit sphere: the relationship between candelas, lumens, steradians, and foot-candles can be shown by using a sphere 1 ft in radius (unit sphere) with a uniform point source of 1 cd at the center of the sphere (see Figure 3-7). For an area of 1 ft² on the surface, the solid angle subtended will be 1 sr.

$$\omega = \frac{dA}{R^2} = \frac{1\ \text{ft}^2}{(1\ \text{ft})^2} = 1\ \text{sr}$$

The 1-cd point source will produce 1 lm in the unit solid angle.

$$d\phi = I d\omega = 1\ \text{cd} \times 1\ \text{sr} = 1\ \text{lm}$$

The illumination produced on the inside surface of the sphere will be 1 lm on 1 ft² or 1 fc.

$$E = \frac{d\phi}{dA} = \frac{1\ \text{lm}}{1\ \text{ft}^2} = 1\ \text{fc}$$

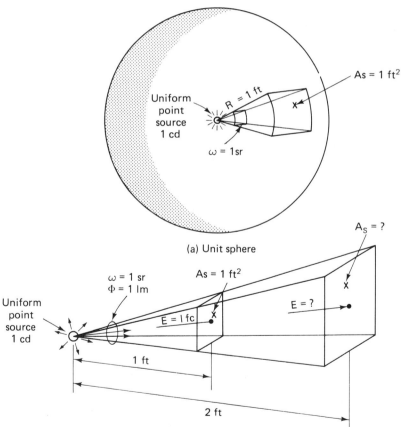

(a) Unit sphere

(b) Segment of the unit sphere

Figure 3-7. Unit Sphere

The total surface area of a sphere is $4\pi R^2$. Therefore, the total surface area of the unit sphere is 4π or 12.57 ft². If the luminous flux of 1 lm falls on each square foot, the 1-cd uniform point source will produce a total of 12.57 or 4π lm.

Mean spherical candlepower: the average candlepower of a light source in all directions in space. It is equal to the total luminous flux of the light source divided by 4π, which is the total surface area of the sphere.

$$\text{MSC} = \frac{\text{total luminous flux}}{\text{total sphere area}} = \frac{\phi_{\text{total}}}{4\pi}$$

Assumptions: the above formula assumes that the light source is uniform in its distribution, which means that the luminous intensity vectors radiating from the point source are the same in all directions. This distribution would be for an indicatrix that is a perfect sphere, which cannot exist in any real light source because of the blocking effect of the lamp base.

The unit is the lumen/steradian, or the candela (cd).

Smashing point: "the time when the lamps have so decreased in light output that it is more economical to break them and replace them with new ones than to continue to burn them."

METRIC SYSTEM

The English system of units will be abandoned in this country in the future. In its place, this country will use the metric system, more properly called the International system of units, abbreviated SI. The primary reasons for adopting the SI units are (1) their extensive use in most countries of the world, (2) they are the primary units in the scientific fields, and (3) the need for uniformity between the scientific and engineering fields.

In illuminating engineering only those terms that involve units of length or areas will be affected by the conversion. Units of lumens, candela, steradians, and efficacy will remain the same. Therefore, only the units of illumination and luminance will be affected by the conversion to SI units.

Illumination (E)

$$1 \text{ fc} = 1 \frac{\text{lm}}{\text{ft}^2} = 10.76 \frac{\text{lm}}{\text{m}^2} = 10.76 \text{ lux}$$

The SI unit of illumination is the lux; the amount of lux is the amount of foot-candles times 10.76. The constant (10.76) is actually the conversion of feet squared to meters squared. The conversion formula is

$$\boxed{\frac{1 \text{ fc}}{10.76 \text{ lux}} = 1}$$

For example,

$$30 \text{ fc} \times \frac{10.76 \text{ lux}}{1 \text{ fc}} = 322.8 \text{ lux}$$

Luminance (L)

$$1 \text{ fL} = \frac{1}{\pi} \frac{\text{cd}}{\text{ft}^2} = 3.426 \frac{\text{cd}}{\text{m}^2}$$

The SI unit of luminance is candelas per square meter. The amount of candelas per square meter is the amount of footlamberts times 3.426. A candela per square meter is equivalent to a unit called a **nits**. The constant (3.426) contains the conversion of feet squared to meters squared and includes the factor of π. The conversion formula is

$$\frac{1 \text{ fL}}{3.426 \text{ (cd/m}^2)} = \boxed{\frac{\text{fL} \times \text{m}^2}{3.426 \text{ cd}} = 1}$$

For example,

$$452 \text{ fL} \times \frac{3.426 \text{ cd}}{\text{fL} \times \text{m}^2} = 1548.6 \frac{\text{cd}}{\text{m}^2}$$

COLOR TEMPERATURE

To define color temperature, the term **blackbody radiator** must be defined. A blackbody is theoretically a complete, perfect radiator. A blackbody radiator changes color as its temperature is raised first to red, then to orange, yellow, bluish white, and white. Color temperature, then, is used to describe the color of a light source by comparing it to the color of a blackbody radiator. For example, the color appearance of an incandescent lamp is similar to a blackbody radiator heated to about 3000 °K. Therefore, it is said that the incandescent lamp has a color temperature of 3000 °K. (The Kelvin scale has a zero point at -273 °C.) The color temperature is not a measure of the actual light-source temperature. It is the temperature of the blackbody radiator when the color appearance is the same. To assign a color temperature to a light source, the light source must match the blackbody radiator in terms of chromaticity (color appearance), as well as have a similar spectral power distribution (SPD) curve. An incandescent lamp has a SPD curve close to that of the blackbody radiator; a chromaticity match is also possible. Therefore, color temperature can be properly assigned to incandescent sources.

Most **white** fluorescent lamps will have a chromaticity match, but they will not have a similar SPD curve to the blackbody radiator. Thus the lamp will not fall on the locus of the blackbody radiator (CIE chromaticity diagram, Chap. 2).

Therefore, it is improper to assign a **color temperature** to a fluorescent lamp. Fluorescent lamps are described in terms of a **correlated color temperature.** A correlated color temperature implies a chromaticity match but not a SPD match, since the source falls off the blackbody locus. Isotemperature lines that cross the locus of the blackbody radiator curve are used to determine the correlated color temperature (see Figure 3-8).

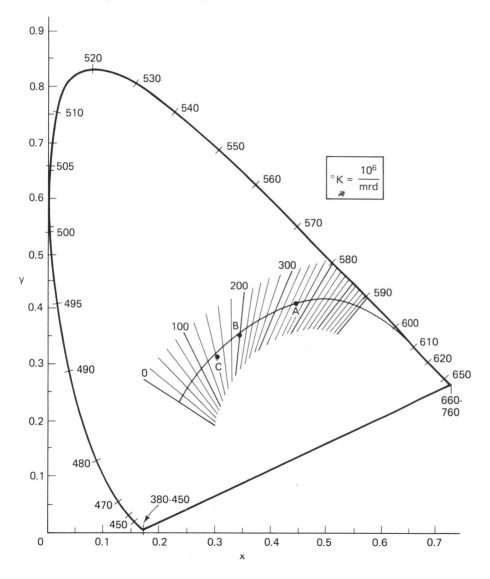

Figure 3-8. Correlated Color Temperature (Isotemperature Lines Plotted on the Blackbody Locus-Values in Microreciprocal Degrees—MRD)

Sources that do not match the blackbody radiator in terms of chromaticity or SPD cannot be properly assigned a color temperature or a correlated color temperature.

PHOTOELECTRIC EFFECT

To understand how illumination measuring devices work, one must understand the basic phenomenon known as the **photoelectric effect.** The photoelectric effect is the liberation of electrons from the surface atoms of a metal. If a metal surface is illuminated by a light quantum, electrons will be liberated from the sur-

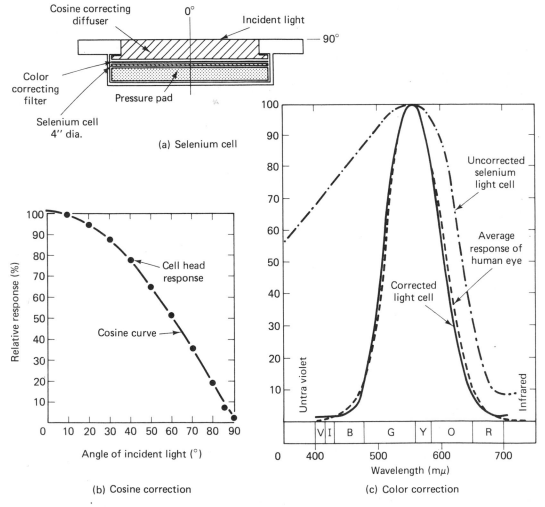

Figure 3-9. Illumination Meter (Courtesy of General Electric Company)

face. If an electric field exists, the electrons will flow to the anode and create an electric current that can be detected by a galvanometer. A standard photocell will generate a small current when illuminated. The photocell consists of a metal plate coated with a semiconductor. When the photocell is exposed to light, electrons are liberated from the metal surface and are trapped at the interface unless an external circuit is provided.

An illumination meter (Figure 3-9) is used to determine the quantity of luminous flux falling on an area (footcandles). Basically, the meter consists of a photocell connected in series with a micrometer calibrated in footcandles. For making interior measurements, the photocell should be cosine and color corrected. The cosine correction is achieved with a specially designed translucent diffusing plate that allows the photocell to correctly evaluate the illumination regardless of direction. The photocell is also color corrected so that its sensitivity matches the standard observer curve (see Chap. 1) or the photopic luminosity curve (V_λ).

reference

1. Kaufman, John, editor, **IES Lighting Handbook,** 5th ed. Illuminating Engineering Society, New York, 1972.

data section

Derivation of Conversion from Luminous Flux (Lumens) to Luminous Intensity (Candelas) (Or, Where Does the π Come From?)

$$dA = 2\pi r \times R d\theta \qquad\qquad \frac{r}{R} = \sin \theta$$

$r = R \sin \theta$, radius of the zone $\qquad r = R \sin \theta$

$dA = 2\pi R \sin \theta \, R d\theta = 2\pi R^2 \sin \theta \, d\theta$, area at an infinitesimal zone surface

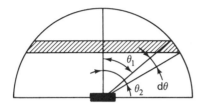 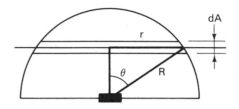

Solid angle subtended: $\quad \omega = \dfrac{dA}{R^2}$, $\quad$ by definition. Therefore,

$$d\omega = d\theta = \frac{2\pi R^2 \sin \theta \, d\theta}{R^2} = 2\pi \sin \theta \, d\theta$$

$$d\phi = I\theta \, d\omega$$

General expression for any source: $\quad \phi = 2\pi \int_0^{\pi/2} I_\theta \sin \theta \, d\theta$

$\qquad\qquad (I_\theta = I_N \cos \theta$, for Lambertian surface$)$

$\qquad\qquad d\phi = I_N \cos \theta \, (2\pi \sin \theta \, d\theta)$

Therefore,

$\qquad\qquad = 2\pi I_N \cos \theta \sin \theta \, d\theta$

$\qquad\qquad \int d\phi = 2\pi I_N \int_0^{\pi/2} \cos \theta \sin \theta \, d\theta$

$\qquad\qquad \phi = 2\pi I_N \left[\frac{1}{2} \sin^2 \theta \right]_0^{\pi/2}$

$\qquad\qquad \phi = \pi I_n$

The amount of lumens is equal to π times the amount of candelas measured normal to the surface. That is, a candela is smaller than a lumen by a factor of π, or 1 cd $= \pi$ lm.

Definition

$$1 \text{fL} = 1 \, \frac{\text{lm}}{\text{ft}^2} \times \frac{\text{cd}}{\pi \, \text{lm}} = \frac{1 \, \text{cd}}{\pi \, \text{ft}^2}$$

Conversion Formulas

$$\frac{cd}{\pi \, lm} = 1$$

TABLE 3-1

Temperature Conversion Equations

To obtain	Conversion equation
°C	$°C = \dfrac{5}{9} \, (°F - 32)$
°F	$°F = \dfrac{9}{5} \, (°C + 32)$
°K	$°K = °C + 273$

TABLE 3-2

Prefixes

Greek prefixes		
Million, mega	1,000,000.	10^6
Thousand, kilo	1,000.	10^3
Hundred, hecto	100.	10^2
Ten, deka	10.	10^1
Latin prefixes		
Tenth, deci	0.1	10^{-1}
Hundredth, centi	0.01	10^{-2}
Thousandth, milli	0.001	10^{-3}
Millionth, micro	0.000 001	10^{-6}

TABLE 3-3

Length Conversion Factors

To Obtain \ Multiply Number of By	Meters	Centimeters	Millimeters	Micrometers	Nanometers (millimicrons)	Angstroms	Feet	Inches
Meters, m	1	0.01	0.001	10^{-6}	10^{-9}	10^{-10}	3.048×10^{-1}	2.540×10^{-2}
Centimeters, cm	100	1	0.1	10^{-4}	10^{-7}	10^{-8}	3.048×10	2.540
Millimeters, mm	1000	10	1	10^{-3}	10^{-6}	10^{-7}	3.048×10^{2}	2.540×10
Micrometers, um	10^{6}	10^{4}	10^{3}	1	0.001	0.0001	3.048×10^{5}	2.540×10^{4}
Nanometers, nm (millimicrons)	10^{9}	10^{7}	10^{6}	10^{3}	1	0.1	3.048×10^{8}	2.540×10^{7}
Angstroms, Å	10^{10}	10^{8}	10^{7}	10^{4}	10	1	3.048×10^{9}	2.540×10^{8}
Feet, ft	3.281	3.281×10^{-2}	3.281×10^{-3}	3.281×10^{-6}	3.281×10^{-9}	3.281×10^{-10}	1	8.333×10^{-2}
Inches, in	3.937×10	3.937×10^{-1}	3.937×10^{-2}	3.937×10^{-5}	3.937×10^{-8}	3.937×10^{-9}	12	1

TABLE 3-4

Area Conversion Factors

To Obtain \ Multiply Number of ... By	Square inches	Square feet	Square yards	Square centimeters	Square meters
Square inches, in.²	1	144	1296	0.155	1550
Square feet, ft²	6.944×10^{-3}	1	9	1.076×10^{-3}	10.76
Square yards, yd²	7.716×10^{-4}	0.1111	1	1.196×10^{-4}	1.196
Square centimeters, cm²	6.452	929	8361	1	10^{4}
Square meters, m²	6.452×10^{-4}	9.920×10^{-2}	0.8361	10^{-4}	1

TABLE 3-5

Luminance Conversion Factors

To Obtain \ Multiply Number of ... By	Candela/in.²	Candela/cm²	Candela/m²	Lambert	Millilambert	Footlambert
Candela/in.²	1	6	6.45×10^{-4}	2.05×10^{-6}	2.05×10^{-3}	2.21×10^{-3}
Candela/cm²	0.155	1	10^{-4}	3.183×10^{-7}	3.183×10^{-4}	3.426×10^{-4}
Candela/m²	1550	10^{4}	1	3.183×10^{-3}	3.183	3.426
Lamberts	0.487	3142	3.142×10^{-4}	1	10^{-3}	1.076×10^{-3}
Millilamberts	487	3142	0.3142	10^{3}	1	1.076
Footlamberts	452	2919	0.2919	0.929×10^{-3}	0.929	1

4 light sources

HISTORY OF THE ELECTRIC LAMP

The success of the electric lamp was dependent on the development of an economical source of electricity. Therefore, the history of the electric lamp is the history of electricity.

About 25 centuries ago a Greek philosopher, Thales, found that a fossil resin called amber became electrically charged when rubbed. The Greek name for amber is **elektron**, which is the basis for the word electricity. Numerous investigators developed crude devices for generating electricity. In 1745, E. G. von

Kleist invented the Leyden jar. The Leyden jar is the forerunner of the condenser. He showed that he could store small quantities of electricity generated by friction machines. Benjamin Franklin connected several Leyden jars in a parallel arrangement (parallel circuits) that produced a strong electrical discharge. His numerous experiments led to the establishment of the principles of parallel and series connections.

Based on Galvani's theory of animal electricity, Alessandro Volta developed the voltaic pile. The voltaic pile consisted of a pile of alternate silver and zinc discs with a piece of cloth soaked in salt water between each metal layer. He showed that electricity could be produced with this arrangement. This was the beginning of the battery. In 1820, a physicist named Hans Oersted discovered that a compass needle could be deflected by current flowing in a wire. Based on Oersted's original findings, André Marie Ampere developed fundamental laws related to current flow in a wire. Ampere showed that current flowing in a wire coil produced a magnetic field, which established a connection between magnetism and electricity.

In 1825, Georg Simon Ohm conducted his important experiments in electricity, which led to the law that states that current flow is directly proportional to the voltage and inversely proportional to the resistance; this is called Ohm's law ($I = E/R$). The invention of the galvanometer by Johann Schweigger made it possible to accurately measure the current flow in a wire. In 1825, William Sturgeon developed the first electromagnet. This was an important step in the process of developing an electric power source that used the principles of magnetism to produce electricity.

In 1831, Michael Faraday, showed that a copper disc rotated between the poles of a permanent magnet would produce an electric current that could be detected on a galvanometer. This was the beginning of many years of investigation of the dynamo (electric generator), which is the foundation of the electric lighting industry.

Prior to 1878, all methods of distributing electric current involved constant-current dynamos (series circuits). This method was satisfactory for arc lamps since they are constant-current devices. However, a series system would be an impractical one for distributing power to electric lamps since it would be impossible to turn off one lamp without turning off all the lamps on the circuit. Thomas Edison knew that if a practical lamp were to be developed a new distribution system would have to be developed to allow individual control of each lamp. Edison developed a constant-voltage dynamo that allowed his incandescent lamps to be operated in parallel. He called this system of distribution a "multiple" system. With the invention of the constant-voltage dynamo, Edison began work on a new high-resistance incandescent lamp. In 1880, Edison was granted a patent on a new, high-resistance, carbon-filament vacuum lamp. Edison demonstrated his new incandescent lighting system at Menlo Park, New Jersey. Although Edison did not invent the first incandescent lamp, he did invent the first practical incandescent lamp and power system that was suitable for multiple distribution over large areas.

The first attempt at making an incandescent lamp was made by Warren de la Rue in 1820. The lamp consisted of a glass tube with brass end cap and a coil of platinum wire for a filament. The lamp had a very short burning life, and the operating cost was very high because of the number of batteries required. Between 1820 and 1878 numerous types, sizes, and configurations of incandescent sources were developed and patented.

The first commercial application of the incandescent lamp was on the steamship **Columbia.** Four dynamos were started on May 2, 1880, and operated 115 lamps on the **Columbia.** The first land application of Edison's system was in the lithography shop of Hinds, Ketchum, and Company in New York City in 1881. Over 150 systems were installed in the next two years utilizing over 30,000 incandescent lamps. The world's largest incandescent electric lamp was built in 1954 to commemorate the seventy-fifth anniversaty of Edison's invention. This lamp, on display in Cleveland, stands 42 inches high and uses 75,000 watts of electricity. It weighs 50 pounds and produces as much light as 2875 sixty-watt light bulbs.

LIGHT SOURCES

Light sources (lamps) used today in artificial lighting can be divided into two main categories: incandescent and gaseous discharge. The gaseous discharge type of lamp is either low or high pressure. Low-pressure gaseous discharge sources are the fluorescent and low-pressure sodium lamps. Mercury vapor, metal halide, and high-pressure sodium lamps are considered high-pressure gaseous discharge sources.

These are the most common light sources used in the field of illuminating engineering. Each light source will be described in terms of its three primary components: (1) light-producing element, (2) enclosure (bulb), and (3) electrical connection. The chapter is divided into three sections: (1) incandescent sources, (2) gaseous discharge sources, and (3) ballasts.

INCANDESCENT SOURCES

Standard Incandescent Lamps

Light-Producing Element

Light is produced in the incandescent lamp (Figure 4-1) by heating a wire or filament to high temperatures, which causes the wire to incandesce (glow). The incandescence of the wire is a result of its resistance to the flow of electrical current through the wire. Tungsten is used as a filament material. No other sub-

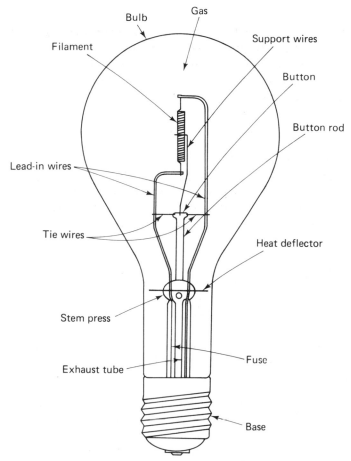

Figure 4-1. Incandescent Lamp (Courtesy of GTE Sylvania Incorporated)

stance is as efficient in converting electrical energy into light on the basis of life and cost. Tungsten has four important characteristics:

1. High melting point.
2. Low evaporation.
3. High strength and ductility.
4. Favorable radiation characteristics.

The most common filament letter designations are s, straight; c, coiled; cc, coiled coil; and r, ribbon or flat. Coiled coil filaments are the most efficient and widely used in the lamps encountered in illuminating engineering. The resistance of cold tungsten is low compared to its operational resistance; therefore, there is a large initial inrush of a current to a cold lamp.

Enclosure

The bulb or glass envelope is used to keep air away from the filament. When the filament is exposed to air, evaporation occurs more rapidly. The bulb is filled with an inert gas of argon and nitrogen to retard evaporation of the filament. Gas-filled lamps designated type C are 40 W and higher. Lamps of 25 W and less are vacuum lamps, which are designated type B. Bulbs are also designated according to their shape (Figure 4-2).

In addition to a letter designation, bulbs also have a number designation, which represents the diameter of the bulb in eighths of an inch. For example, an A—19 designation indicates a diameter of $^{19}\!/\!_8$ or $2\frac{3}{8}$ in.

Bulb surfaces treatments may be clear, surface treated, tinted, or built-in silvered control surfaces. The standard lamps on the market are clear, inside frosted (or acid etched), white (silica coating), and white bowl or silver bowl. Colored or tinted bulbs are natural-colored glass, outside spray, ceramic glazing (enameled), or fused-on color filters.

Electrical Connection

The base provides an electrical connection, mounting, and positioning of the lamp. There are eight types of bases: (1) screw, (2) screw with ring contacts (three-way), (3) skirted screw, (4) bipost, (5) prefocus, (6) disc, (7) bayonet, and (8) prong.

The general-service lamps of 300 W and below normally use the medium screw base; 300- to 500-W lamps use the mogul screw base.

Operating Characteristics

VOLTAGE FLUCTUATION. Varying the voltage of an incandescent lamp above or below the rated voltage will affect the characteristics of the lamp. For example, if a 120-V-rated lamp is operated at 125 V (104 percent increase), the lamp will produce 16 percent more lumens, 7 percent more watts, and 38 percent less life. A 120-V-rated lamp operated at 115 V (96 percent) will provide 13 percent less lumens, 6 percent less watts, and 62 percent more life (see Figure 4-3).

LUMEN DEPRECIATION. The resistance of the filament increases with time owing to evaporation and accompanying decrease in the diameter of the filament. This increase in filament resistance causes a decrease in lumens, amperes, and watts. Further reduction in lumen output is due to the absorption of light by the deposited tungsten on the inside surface of the lamp (see Figure 4-4).

Tungsten–Halogen Lamp

A deficiency of standard incandescent lamps has been lumen maintenance over life. When the filament is heated, it slowly evaporates and redeposits on the inside of the bulb wall. This layer of tungsten than acts like a filter, absorbing some light and lowering the light output. This was overcome by the development of the *tungsten–halogen* cycle lamp, which is also called a quartz lamp. The tungsten–halogen lamp contains a halogen such as iodine or bromine and a fill

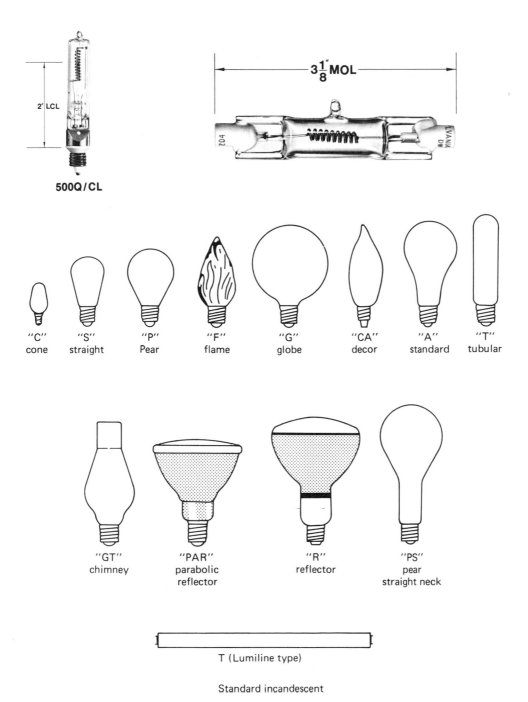

2' LCL

3$\frac{1}{8}$" MOL

500Q/CL

"C"
cone

"S"
straight

"P"
Pear

"F"
flame

"G"
globe

"CA"
decor

"A"
standard

"T"
tubular

"GT"
chimney

"PAR"
parabolic
reflector

"R"
reflector

"PS"
pear
straight neck

T (Lumiline type)

Standard incandescent

Figure 4-2. Standard Incandescent and Tungsten-Halogen
Bulb Shapes (Courtesy of GTE Sylvania, Inc.)

73

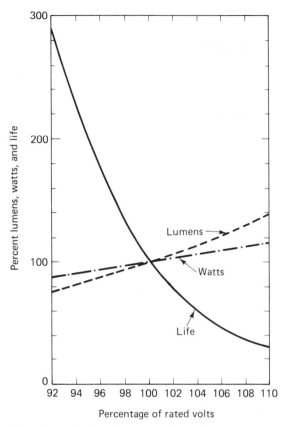

Figure 4-3. Voltage Effect on Lamp Life and Lumens Output
(Courtesy of GTE Sylvania Incorporated)

gas. The bulb or envelope is made of quartz to withstand the high temperatures required for the cycle to work. At high temperatures, the evaporated tungsten associates with a halogen molecule. Instead of being deposited on the bulb wall, the combined tungsten–halogen molecule is returned to the hot filament, freeing the halogen to pick up another evaporated tungsten molecule. This cleaning action minimizes the deposit of tungsten on the bulb wall, and results in an increased lumen output throughout the life of the lamp. Figure 4-4 shows the lumen output of a standard incandescent lamp and that of a tungsten–halogen lamp over the life of each.

The main objective in developing the tungsten–halogen lamp was to maintain lumen output, but other improvements were also realized. Lamp life increased slightly, as did efficacy. To operate properly, tungsten–halogen lamps require relatively high temperatures. To obtain these high temperatures, the filament had to be compacted and the outer envelope made smaller. The smaller source more closely approaches the ideal point source needed for good optical control.

The tungsten-halogen is a type of incandescent lamp and is thus easily dim-

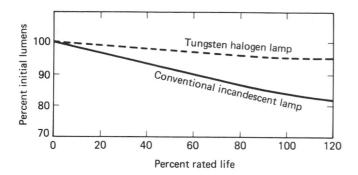

Figure 4-4. Lumen Depreciation of Tungsten-Halogen and Conventional Lamp (Courtesy of GTE Sylvania Incorporated)

med. Dimming, however, causes a reduction in bulb wall temperature, which retards the uniting of the tungsten and halogen molecules, resulting in bulb wall blackening and a reduction in lumen output. When the lamp is returned to a sufficient heat level, some of the tungsten deposited on the bulb is removed.

General Operating Characteristics

Efficacy and Life

One of the most important characteristics of any light source is its ability to convert electrical energy into luminous energy. This is known as **lamp efficacy.** The incandescent lamp has efficacies ranging from 4 to 24 lm/W. For comparison purposes, the incandescent lamp is typically said to have an efficacy of 20 lm/W.

The cost of light depends not only on the efficacy but also on the life of the source. Incandescent lamps have an average life of 1000 h or about 5 months for a typical burning period of 8 h/day (52 wk/yr × 6 days/wk × 8 h/day = 2496 h/yr). Lamp life is a function of many factors, including filament configuration and support, fill gas, on–off cycles, and wattage.

Color Characteristics

The human visual system responds differently to the different wavelengths of radiation. Our mind interprets these different wavelengths as color. Light sources are important in color vision because they provide radiant energy and thus color response. The distribution of wavelengths emitted by a source is known as the **spectral power distribution** (SPD). The significance of the SPD curve and its relationship to color and color vision are discussed in Chap. 2. The SPD of an incandescent lamp is shown in Figure 4-5. Note the tremendous amount of red or long wavelengths present; this is to be expected from a source producing luminous energy by heat. The quartz iodide SPD is similar to that of the incandescent lamp but contains slightly more of the shorter wavelengths (blue). This is the result of the higher operating temperatures. Incandescent lamps have acceptable color rendition.

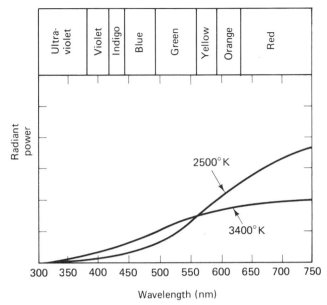

Figure 4-5. Spectral Power Distribution for an Incandescent
Lamp (Courtesy of GTE Sylvania Incorporated)

Energy Distribution.   Energy distribution of an incandescent lamp is illustrated
in Figure 4-6.

Summary

Although the incandescent source has a short life and low efficacy, it has
advantages that make it a common choice for a light source. Among the advantages is the low initial cost of a lamp. Also, the relatively small physical size
makes it easy to direct the light output, because it approaches the ideal model of a
point source. Color rendition is acceptable. Often an incandescent system is
chosen because it is the easiest and cheapest to dim, an important consideration in
many designs.

The energy-saving lamps on the market now make use of different fill gases.
These lamps use krypton rather than the argon gas used in conventional lamps.
The result is a decrease in wattage without a decrease in efficacy. As an added
benefit, the life is increased. The incandescent lamp is still popular because of its
low cost. Energy-saving lamps are about 10 times the initial cost of an equivalent
conventional incandescent lamp.

GASEOUS DISCHARGE SOURCES

Gaseous discharge lamps are referred to as zero-resistance elements or
negative-resistance elements. As the additives inside the arc tube ionize, the

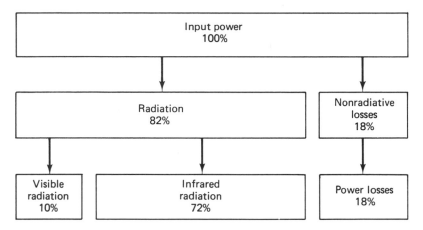

Figure 4-6. Energy Distribution of an Incandescent Lamp.
17.5 lm/watt (100W) @ 10% visible radiation

resistance inside the arc tube decreases. This will cause the resistance to approach zero while the current approaches infinity.

$$I = \frac{E}{R} \qquad R \to 0, \quad I \to \infty$$

Without a device to limit the current, the electrodes would burn up within a matter of seconds. Therefore, *all* gaseous discharge sources require a ballast. A ballast is an electrical device that serves three primary functions:

1. Limits the current (zero-resistance element characteristics).
2. Provides a starting voltage kick.
3. Provides power factor correction.

The ballast acts as an autotransformer to provide the voltage kick. Therefore, it contains windings that will cause an inductive reactance load (see Chap. 8). Inductive reactance causes a phase shift between the current and voltage, which is corrected by the addition of a capacitor to the ballast. Ballast will be described in more detail at the end of this chapter.

Burning Position

Gaseous discharge lamps are usually sensitive to burning position. The engineer must use caution in matching the lamps selected with the final burning position of the lamp. If lamps are burned in a position other than intended, the lamp life, lumen output, and color characteristics can change. Some lamps may explode or implode if not properly installed. Manufacturer specifications should be consulted for burning position information. Typical letter designations for burning position are

BU: base up   BD–HOR: base down to horizontal
BD: base down   VER–BU: vertical to base up
BU–HOR: base up to horizontal   VER–BD: vertical to base down
                                                      HOR: horizontal only

LOW-PRESSURE GASEOUS DISCHARGE

Fluorescent Lamps

The first major showing of the fluorescent lamp was at the 1938–1939 New York World's Fair. The lamps were installed in vertical clusters on the flag poles along the Avenue of Flags. Figure 4-7 is a schematic showing the operation of a fluorescent lamp.

Light-Producing Elements

The fluorescent lamp requires three elements or components to produce visible light: (1) electrodes, (2) gases, and (3) phosphor.

ELECTRODES (CATHODES).   Electrodes are the current-emitting devices. Two types of cathodes are in use today. The **hot cathode** is a coiled coil or a triple-coiled tungsten filament coated with an alkaline earth oxide that emits electrons when heated. The electrons are boiled off at about 900 °C. The **cold-cathode** lamp is a pure iron tube that also has an electron-emitting material applied inside the tube. The electrodes are subjected to higher voltage, releasing electrons at about 150 °C. The hot cathode is the most common type of electrode used in fluorescent lamps for most applications. Therefore, we shall not describe cold-cathode lamps.

GASES.   A small quantity of mercury droplets is placed in the fluorescent tube. During the operation of the lamp, the mercury vaporizes at a very low pressure. At this low pressure, the current flowing through the vapor causes the vapor to radiate energy principally at a single wavelength in the ultraviolet (253.7 nm) region of the spectrum. The pressure of the mercury is regulated during operation by the temperature of the bulb wall.

The lamp also contains a small amount of a highly purified rare gas. Argon and argon–neon are the most common, but krypton is also sometimes used. The gas ionizes readily when a sufficient voltage is applied to the lamp. The ionized gas decreases in resistance quickly, allowing current to flow and the mercury to vaporize.

PHOSPHOR.   This is the chemical coating on the inside wall of the bulb. When the phosphor is excited by ultraviolet radiation at 253.7 nm, the phosphor produces

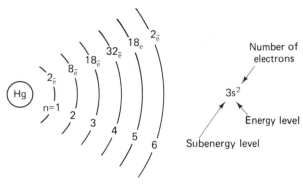

$$Hg(1s^2, 2s^2, 2p^6, 3s^2, 3p^6, 4s^2, 4p^6, 4d^{10}, 4f^{14}, 5s^2, 5p^6, 5d^{10}, 6s^2)$$

(a)   Mercury atom

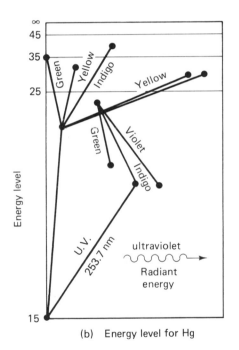

(b)   Energy level for Hg

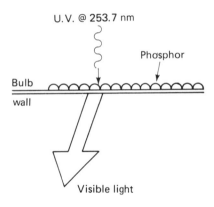

(c)   Fluorescence - U.V. radiant energy excites phosphor to produce visible light

Figure 4-7.   Schematic of the Operation of a Fluorescent Lamp

visible light by fluorescence. (See Figure 4-7.) That is, visible light from a fluorescent lamp is produced by the action of ultraviolet energy on the phosphor coating on the inside of the bulb.

Enclosure

The bulb is the glass envelope that holds the gases in the tube and also provides a surface to which the phosphor coating can be applied. Bulbs are designated according to shape, diameter, and color (see Figure 4-8). For example, T-12 would indicate a tubular shape (T), and a diameter of 1½ in. (12 represents the diameter in eighths of an inch: 12/8 = 1½ in.).

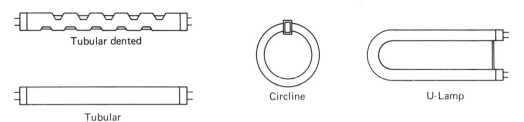

Tubular dented

Tubular

Circline

U-Lamp

Figure 4-8. Fluorescent Lamp Shapes (Courtesy of General Electric Company)

Electrical Connection

The base forms the electrical connection of the lamp to the socket and acts to support and align the lamp. Three basic types of bases are associated with the fluorescent lamp:

1. Bipin (miniature, medium, mogul): used on all preheat and most rapid start lamps.

2. Bipin recessed double contact: used on the high-output and power groove lamps. Its purpose is to protect the user from the high-voltage pin contacts.

3. Single pin: used for instant-start lamps.

Color Characteristics

The color or SPD of a fluorescent lamp depends on the phosphor coating on the inside of the bulb wall. The spectral power distribution curve consists of two components: (1) a continuous or smooth portion, and (2) a line spectrum. The lines or bars on the SPD curve represent visible light that is generated directly by the mercury arc; the smooth portion is due to the action of the ultraviolet energy on the phosphor. The SPD of a fluorescent lamp can be changed by modifying the type and mix of phosphors used to coat the lamps. There are six standard white fluorescent lamps on the market (see Figure 4-9):

CW: cool white	CWX: cool white deluxe
WW: warm white	WWX: warm white deluxe
W: white	D: daylight

This assortment of white fluorescent lamps has been developed to meet nearly all needs for white light. They are referred to as **standard white** lamps because all six are available from every major manufacturer of fluorescent lamps. In addition to these six standard whites, each manufacturer markets special whites and colored fluorescent tubes.

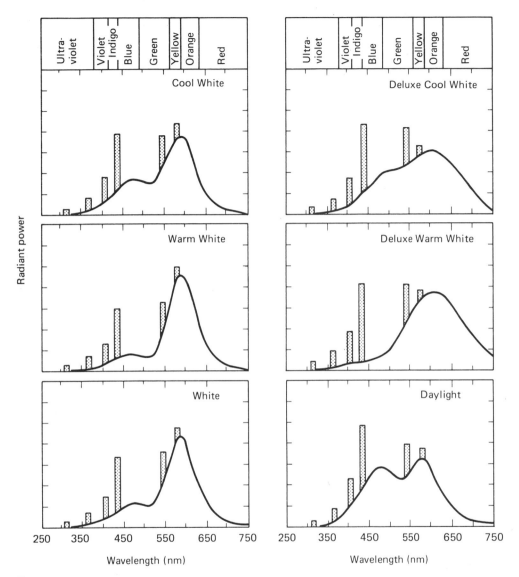

Figure 4-9. Spectral Power Distribution for the Standard Fluorescent Lamps (Courtesy of General Electric Company)

A choice among fluorescent lamps always means a compromise between efficacy and color. Selection of the best color rendition usually means a reduction in efficacy. The CW, WW, W, and D lamps have higher efficacy, but they are weak in red, resulting in poor color rendition characteristics. The CWX and WWX are designed to give the best color rendition to objects and people at a reasonable efficacy. This is achieved by the addition of red phosphors to the mixture. However, since the eye is less responsive to red energy, the luminous efficiency is reduced by about 30 percent of the light output of the CW and WW lamps.

Hot-Cathode Circuits.   There are three types of hot cathode fluorescent lamps. They are defined by the circuits for which they are designed:

1. Preheat
2. Instant start
3. Rapid start

Preheat Circuit.   The preheat circuit was the first type to be developed. It requires a separate starter that preheats the electrodes, causing an emission of electrons. This causes the internal resistance to drop, which allows the arc to strike. The preheating process requires a few seconds, hence the slow, blinking start that is characteristic of the preheat circuit. Preheating may be accomplished by a manual button starter or an automatic starter. The starter applies current to the electrodes of the lamp for a sufficient time to heat them and then automatically (or manually) removes the current from the electrode, causing the voltage to be applied between the electrode and striking the arc.

Instant-Start Circuit.   In 1944, the instant-start circuit was introduced to overcome the slow starting characteristics of the preheat circuit. The instant-start circuit eliminates the need for a starter and thus simplifies the system and maintenance. A sufficiently high voltage is applied between the electrodes to break down the resistance of the lamp and strike the arc. The arc quickly heats up the filament electrodes, which then supply electrons to sustain the arc. Since no preheating is required with instant-start lamps, a single-pin contact is sufficient. This type of lamp is called a **slimline** lamp.

Rapid-Start Circuit.   In 1952, the rapid-start circuit and lamp were developed. It starts quickly without the need for a separate starter. A rapid-start ballast is smaller and more efficient than the instant-start ballast for the same wattage. The rapid-start circuit utilizes low-resistance electrodes that are heated continuously with very low losses.

The rapid-start lamp is the most common lamp and is best suited for most applications. Rapid-start circuits can be flashed and dimmed quite efficiently.

Circline (circular) lamps are available for operation in rapid-start circuits. Two-foot hair-pin or U-lamp fluorescents also are designed for use in rapid-start circuits. Rapid-start lamps can be used in both preheat and rapid-start circuits. However, a lamp carrying the designation "Preheat" cannot be used in a rapid-start circuit. Rapid-start circuits are classified according to the lamp input current rating:

RS	430 mA
Circline	390, 420, 430 mA
HO	800 mA
XHO, PG, VHO, SHO, T10	1500 mA

Lamp Designation

A lamp is designated according to the wattage or length, shape, diameter in eighths of an inch, and color. The preheat and rapid-start lamps use the nominal lamp wattage in their designation while the high output, very high output, instant start, and power groove lamps use the nominal length in their designations. Some examples follow:

Preheat:
　F20T12/CW, fluorescent/watts/tubular/diameter/color
Rapid start, 30 and 40 watt:
　F30CW and F40CW, fluorescent/watts/color
Rapid start (high output):
　F96T12/CW/HO, fluorescent/length/tubular/diameter/color/high output
Instant start:
　F96T12/CW/SL, fluorescent/length/tubular/diameter/color/slimline
Rapid start, power groove:
　F96PG17/CW, fluorescent/length/power groove/diameter/color
Rapid start, circline:
　FC16T10/CW,RS, fluorescent/diameter of circle/tubular/diameter/color/rapid start

Performance Characteristics

LIFE.　Lamp life is dependent on the burning cycle per start. Lamp ratings are given on the basis of a 3-h burning cycle per start. In 1973 a new collector gas was introduced into lamps. This gas prevents the boiling off of the electron emission material from the electrodes every time that the lamp is started; therefore, the life of the lamp is no longer affected by more frequent on–off cycling of the lamp. However, the importance of burning time can be seen in the extended operation of lamps in terms of the life-multiplier factors:

$$6\text{-h burning/start: } 1.25 \times \text{life}$$
$$12\text{-h burning/start: } 1.60 \times \text{life}$$
$$\text{continuous burning: } 2.5 \text{ or more} \times \text{life}$$

STROBOSCOPIC EFFECT.　Stroboscopic is a Greek word that means "to see motion." The arc stream extinguishes during each reversal (120 times per second) of the ac sine wave; however, the phosphor coating continues to radiate light during this brief period. Generally, this is not noticeable, but it can make high-speed machinery appear to stand still. The use of a series sequence ballast on rapid-start circuits will eliminate this problem. Another solution is to use a lead-lag ballast, which puts one lamp out of phase with the other in a two-lamp unit. This results in one lamp being at maximum light output while the other lamp is at zero output. The net effect is to eliminate the flicker.

EFFECT OF TEMPERATURE.　The most efficient lamp operation is achieved when the ambient temperature is between 70 to 90 °F. Lower temperatures cause a reduction in mercury pressure, which means that less ultraviolet energy is produced; therefore, less UV energy is available to act on the phosphor and less light is produced. High temperature causes a shift in the wavelength produced that is nearer to the visual spectrum. The longer wavelengths have less effect on the phosphor, and therefore less light output (see Figure 4-10).

Standard fluorescent lamps can be operated down to a temperature of 50 °F without special ballasting. However, as Figure 4-10 indicates, the light output (lumens) will be less if the ambient temperature is outside the 70 to 90 °F range. Special low-temperature ballasts are available for starting and operating lamps at 0 and − 20 °F. These ballasts provide a higher starting voltage, and they usually contain a thermal starting switch.

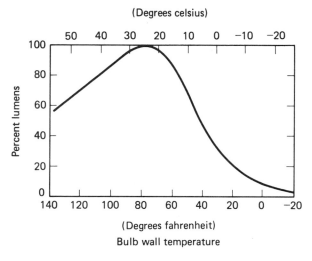

Figure 4-10. Temperature Effects on Fluorescent Lamps (Courtesy of General Electric Company)

EFFECT OF HUMIDITY.   Starting voltage requirements are affected by the electrostatic charge on the outside of a fluorescent lamp. Moist, humid air has unfavorable effects on the surface charge. This factor must be taken into account when the relative humidity exceeds 65 percent. A silicone coating on the outside surface of the lamp and the proper distance between the lamp and housing will usually solve starting problems under any conditions of humidity. However, dirt accumulation on the lamp will nullify the effects of the silicone coating and cause starting difficulties. Cleaning the lamp with an abrasive cleaner may also remove the silicone coating.

ENERGY DISTRIBUTION.   Of the total input energy into the fluorescent lamp, only 22 percent ends up as visible light (see Figure 4-11).

EFFICACY.   Fluorescent lamp efficacy for the most common sizes of lamps

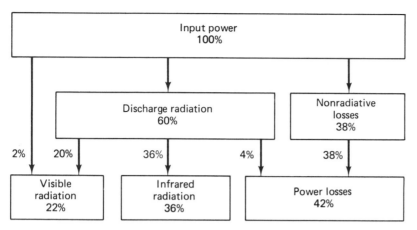

Figure 4-11. Energy Distribution for a Fluorescent Lamp
78.8 lm/watt (40W) @ 22% visible radiation

ranges from 75 to 80 lm/W not including the ballast. For two F40CW lamp circuits, the total efficacy (lamps plus ballast) would be 68.5 lm/W.

$$2F40CW: 2 \times 3150 = 6300 \text{ lm}$$

Two-lamp, high-power-factor, rapid-start ballast = 92 W

$$\text{Efficacy} = \frac{6300}{92} = 68.5 \text{ lm/W}$$

The F40CW lamp alone has an efficacy of 78.8 lm/W.

ENERGY-SAVING DEVICES.   Energy-saving lamps are designed to operate at a lower wattage on the same ballast as conventional fluorescent lamps. The efficacy of some is decreased; others have an increase in efficacy. Recently, it has been discovered that energy-saving lamps may be the cause of premature ballast failure due to capacitor overloading. A high-power-factor two-lamp ballast contains a starting capacitor and a power factor correcting capacitor. An increase of 6 percent in the current to the starting capacitor is the cause of ballast failure. New ballast designs have eliminated the problem; however, older ballasts or defective ballasts may still show a high failure rate. Energy-saving lamps should only be considered for retrofitting an existing installation that was improperly designed and overlighted; they should not be used for new installations.

For economic reasons, the two-lamp fluorescent luminaire was preferred over a single-lamp luminaire prior to the energy crisis. The two-lamp luminaire may have produced lighting levels higher than required, but at the time energy cost was very low. Because of the lower energy cost, it was more economical to buy a luminaire that operated two lamps with a single (two-lamp) ballast than it was to buy a luminaire with a ballast that operated one lamp. The single-lamp ballast costs about the same as a two-lamp ballast, but more light could be produced with less energy with the two-lamp luminaire.

F40CW: 3150 lm/lamp

Single-lamp, high-power-factor ballast = 52 W

$$\text{system efficacy} = \frac{3150}{52} = 60.6 \text{ lm/W}$$

Two-lamp, high-power-factor ballast = 92 W

$$\text{system efficacy} = \frac{6300}{92} = 68.4 \text{ lm/W}$$

With the development of new ballast circuits and continuous cathode heating, the stroboscopic effect associated with single-lamp units should be minimized. With rising utility rates (operating cost) and the emphasis on reduced power consumption, the use of single-lamp luminaires becomes more important. Removing one lamp from a two-lamp luminaire may appear to be a simple solution for reducing energy consumption in an existing building where unnecessarily high illumination levels exist in noncritical task areas. But since the two lamps are wired in series, the removal of one lamp will result in the other lamp going out. This problem has been solved by the development of a dummy tube that replaces one fluorescent tube to complete the series circuit of the two-lamp luminaire. The dummy tube is constructed of glass and contains a capacitor that offsets the ballast inductance. The capacitance restores the system to its normal power factor while allowing the remaining fluorescent to operate. A typical two-lamp, F40, rapid-start circuit will show a 62 percent drop in the wattage consumed when one lamp is replaced with a dummy tube. At the same time, the light output of the two-lamp luminaire will decrease to 67 percent of the original light output. This will result in approximately a 7 percent increase in efficacy.

Two-lamp, high-power-factor ballast = 92 W
   2F40CW rated at 3150 lm = 6300 lm
Dummy lamp:
   92 W × 62%   = 57.04 W
   2 × 3150 × 67% = 4221 lm

$$\text{Efficacy} = \frac{4221}{57.04} = 74 \text{ lm/W}$$

Normal two-lamp operation:

$$\text{Efficacy} = \frac{6300}{92} = 68.5 \text{ lm/W}$$

The use of dummy tubes should be limited to retrofit applications since the dummy tubes are quite expensive. Also, removing a single lamp from a two-lamp luminaire will cause a nonuniform appearance to the surface of the lens.

Low-Pressure Sodium Lamps

The low-pressure sodium lamp has been used extensively in Europe since the 1940s. A major marketing campaign was begun in the United States in 1972. The low-pressure sodium lamp has the highest lamp efficacy of all sources, but it is monochromatic yellow.

Light-Producing Element

The light-producing element is an arc tube. The arc tube is U shaped and constructed of borate glass. The tube is dimpled to maintain a uniform distribution of sodium throughout. The arc tube contains a small amount of argon and neon to aid in starting the lamp. The pressure inside the arc tube is approximately 1×10^{-3} mm; the enclosing space is under a vacuum.

$$\text{starting time} = 9 \text{ min } (89\%), \quad 15 \text{ min } (100\%)$$
$$\text{restrike} \quad = 30 \text{ sec } (80\%)$$

Enclosure (Bulb)

The bulb is made of ordinary glass. It acts to maintain a constant environment for the arc tube. The space between the bulb and arc tube is under a vacuum. The arc tube operates at a temperature of 260°C (500°F).

The lamp comes in five wattages:

Normal watts	MOL	Bulb shape	Burning position
35	$12^3/_{16}$	T17	HOR/UP
55	$15^3/_4$	T17	HOR/UP
90	$20^3/_4$	T21	HOR only
135	$30^1/_2$	T21	HOR only
180	$44^1/_8$	T21	HOR only

Electrical Connection

The base is a bayonet base (BAY-B1) that will maintain the U-shaped arc tube in a horizontal orientation.

Color Characteristics

The light produced by a low-pressure sodium lamp is a monochromatic yellow (see Figure 4-12). The spectral power distribution consists of two lines at 589 nm (approximately 95 percent of the output) and 586 nm (approximately 5 percent of the output). Because of the monochromatic yellow characteristics, the color rendering potential of the lamp is nonexistent. All colors appear as different shades of gray or brown except for yellow objects.

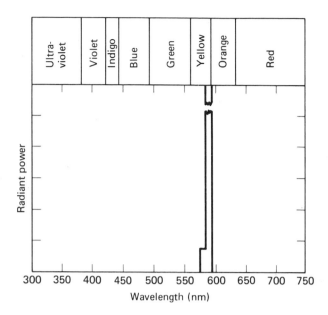

Figure 4-12. Spectral Power Distribution for a Low-Pressure Sodium Lamp

Lamp Designation

The SOX designation is used to indicate a low-pressure sodium lamp. The designation also includes the nominal lamp wattage, such as SOX 180 (180 W).

Performance Characteristics

LUMEN DEPRECIATION.   The lumen output increases slightly over the life of the lamp. Lumen output is said to be constant over an operating temperature range of −10°C (+14°F) to +40°C (+104°F). The effect on lumen output when the lamp is operated outside this temperature range has not been published.

LIFE.   The rated life for all wattages is 18,000 h based on 5 h per start burning cycle. Burning position is critical to lamp life since lamp failure is due to the migration of the sodium toward the electrodes. This migration causes an increase in the watts consumed by the lamp over its life, which results in electrode failure.

ENERGY DISTRIBUTION.   Figure 4-13 shows the energy distribution for a low-pressure sodium lamp.

EFFICACY (LAMP ONLY).   For the effects of ballast losses on efficacy, refer to the discussion on ballasts in this chapter.

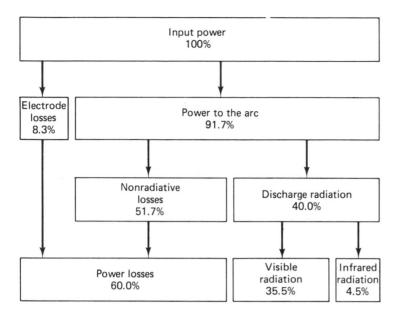

Figure 4-13. Energy Distribution of a Low-Pressure Sodium Lamp 183 lm/watt (180W) @ 35.5% visible radiation.

Nominal watts	Lumens	Actual lamp watts (100 h)	Lamp efficacy (100 h)	Actual lamp watts (18000 h)	Lamp efficacy (18000 h)
35	4,640	36	129.2	44	105.7
55	7,700	53	145.3	62	124.2
90	12,500	90	138.9	122	102.5
135	21,500	130	165.4	178	120.8
180	33,000	176	187.5	241	136.9

HIGH-PRESSURE DISCHARGE SOURCES (HIGH-INTENSITY DISCHARGE SOURCES)

Mercury Vapor Lamps

Light-Producing Element

The light-producing element is an arc tube. The arc tube is constructed of quartz to allow ultraviolet radiation to be transmitted (see Figure 4-14). The arc tube contains mercury and small quantities of argon, neon, and krypton. When the lamp is energized, an arc is struck between the main and starting electrode. As

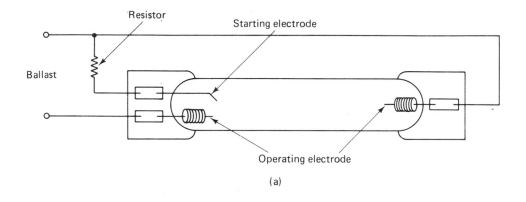

Resistor

Starting electrode

Ballast

Operating electrode

(a)

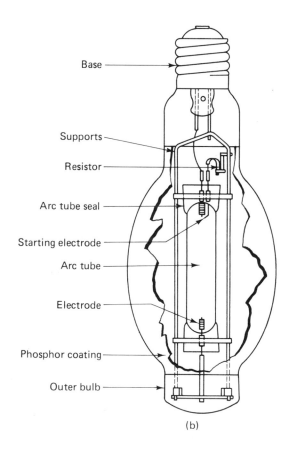

Base

Supports

Resistor

Arc tube seal

Starting electrode

Arc tube

Electrode

Phosphor coating

Outer bulb

(b)

Figure 4-14. Mercury Vapor Lamp and Arc Tube (Courtesy of Westinghouse Lamp Division)

the mercury ionizes, resistance inside the arc tube decreases. When resistance inside the arc tube is less than external resistance, the arc jumps between the main electrodes. The mercury continues to ionize, increasing the light output. The light produced is in the typical mercury lines (404.7, 435.8, 546.1, 577.9), plus UV energy. The arc tube is operated at from 1 to 10 atmospheres of pressure.

starting time = 5 min (80%),   7-10 min (100%)
restrike time = 7 min (80%)

Enclosure (Bulb)

The outer enclosure serves three primary functions:

1. Ordinary glass acts as a UV filter, which prevents skin and corneal burns.
2. It provides a constant thermal environment for the arc tube. The arc tube pressure is affected by rapid temperature change and air motion.
3. It provides a surface for a phosphor coating, which is placed on the inside of the outer enclosure to correct the color rendition of a mercury vapor lamp. A phosphor-coated lamp will require very large housing for good optical control since the coated outer bulb becomes the light source.

Electrical Connection

A mogul screw base is used for all higher-wattage lamps; the 40/50-, 75-, and 100-W lamps come in a medium screw base.

Color Characteristics

A clear mercury vapor lamp has a predominate blue-green color characteristic associated with the mercury line spectrum. Figure 4-15 gives the SPD curve. To color correct the lamp, a phosphor coating is placed on the inside surface of the outer bulb. The primary colors added by the phosphor are the reds and oranges. Phosphor-coated or white mercury vapor lamps are recommended for all applications where color is important. There are three standard white mercury vapor lamps:

1. Color improved: very poor on reds, marginal color, not recommended.
2. Deluxe white, DX: increased red, good color, recommended.
3. Warm white deluxe, WWX: excellent reds, excellent color, highly recommended, decreased lumens.

Lamp Designation

The lamp designation for mercury vapor lamps is quite different from incandescent or fluorescent lamps. The only parts that have meaning are the H designation, which identifies the lamp as a mercury vapor (Hg, mercury), and the wattage. The remaining number and letter are arbitrary.

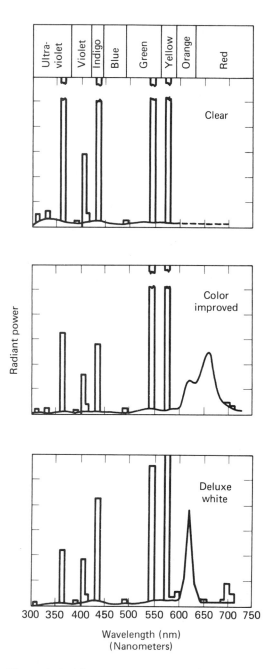

Figure 4-15. Spectral Power Distribution Curve for Mercury-Vapor Lamps (Courtesy of Westinghouse Lamp Division)

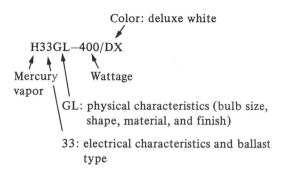

The bulb is designated in terms of a letter, number combination. The letter or letters are used to designate the bulb shape (see Figure 4-16):

PAR: parabolic   BT: bulged tubular
PS: pear straight   R: reflector
T: tubular   E: elliptical
B: bulged   A: standard

The number represents the maximum diameter of the lamp in eighths of an inch.

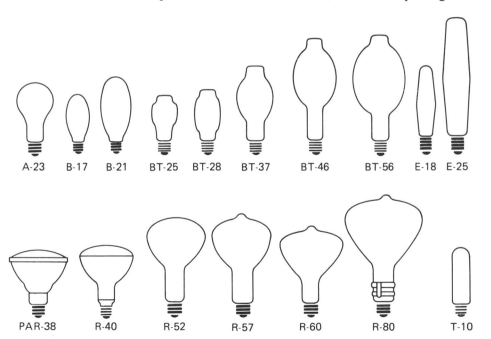

Figure 4-16. HID Bulb-Shape Designations (Courtesy of Westinghouse Lamp Division)

BT-37

$$\text{Diameter} = \frac{37"}{8} = 4\frac{5"}{8}$$

Shape: bulged tubular

Burning position is a function of the location of the starting electrode. The starting electrode should always be located at the top of the lamp to prevent mercury from collecting at the starting electrode, which will adversely affect lamp life.

Performance Characteristics

LUMEN DEPRECIATION.   The lumen depreciation curve for a mercury vapor lamp is quite drastic and a function of the ballast and wattage (see Figure 4-17). The mercury vapor lamp is the only lamp that is listed with a **rated** and **usable** life. The usable life is well suited to the definition of "smashing point" (see Chap. 3). Light output is also a function of the supply and regulation of the voltage to the lamp (see Figure 4-18).

LIFE.   The life of a mercury vapor lamp can be described in terms of its usable or its rated life. Typically, lamps are rated on the basis of a 50 percent mortality curve. Because of rapid lumen depreciation, the mercury vapor lamp is rated in

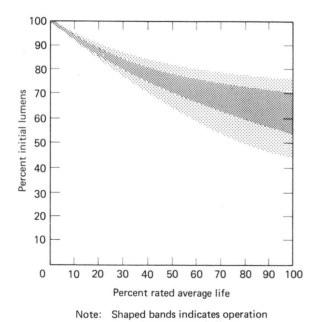

Note: Shaped bands indicates operation
on various mercury ballasts

Figure 4-17. Lumen Depreciation for Mercury-Vapor Lamps
(Courtesy of Westinghouse Lamp Division)

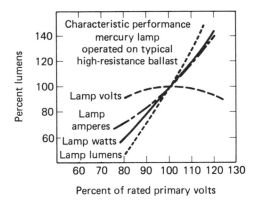

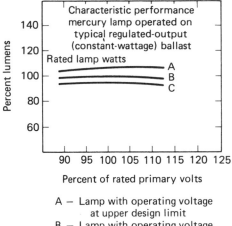

A — Lamp with operating voltage
at upper design limit
B — Lamp with operating voltage
at center value
C — Lamp with operating voltage
at lower design limit

Figure 4-18. Voltage versus Light Output (Courtesy of Westinghouse Lamp Division)

terms of a larger percent of lamps remaining to maintain a more reasonable lumen output (see Figure 4-19).

ENERGY DISTRIBUTION.   The energy distribution for the mercury vapor lamp is shown in Figure 4-20.

LAMP EFFICACIES.   Lamp efficacy varies with lamp wattage. The higher the lamp wattage, the higher the efficacy.

40/50 W: 25 to 30 lm/W
75, 100, 175, 250 W: 34 to 48.4 lm/W
400 W: 55 to 60 lm/W
1000 W: 57 to 63 lm/W

H33GL — 400/DX rated at 22,500 lm

$$\text{Efficacy} = \frac{22,500}{400} = 56.3 \text{ lm/W}$$

SELF-BALLASTED MERCURY.   Self-ballasted mercury lamps contain either a solid-state starting device or an incandescent filament that acts as a ballast. The solid-state type of lamp should not be used in a totally enclosed luminaire because of the heat buildup associated with this lamp.

In general, self-ballasted mercury lamps are 50 percent less efficient than standard mercury lamps, but 50 percent more efficient than incandescent lamps. They should be limited to replacement of incandescent lamps where relamping is difficult and ballast addition impractical.

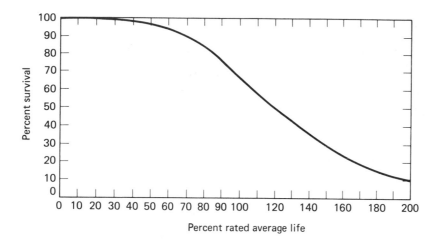

Figure 4-19. Typical Mortality Curve for Mercury Lamps (Courtesy of Westinghouse Lamp Division)

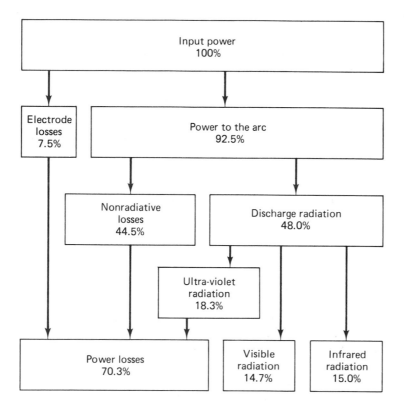

Figure 4-20. Energy Distribution for Mercury Vapor Lamps 56.3 lm/watt (400W) @ visible radiation.

96

ENERGY-SAVING DEVICES.   Recent developments for mercury vapor include dimming capability (see Chap. 8) and electronic ballasting. Electronic ballasts have been researched since the mercury vapor lamp first appeared. A number of problems remain, including high cost, but it is known that with an electronic ballast lamp efficacy and total system efficacy are increased significantly. Other advantages expected of the electronic ballast are smaller size and weight, lower noise, increased lamp life, and easier dimming.

Metal Halide Lamps

Light-Producing Element

The light-producing element is an arc tube. The arc tube has the same construction and operating principles as the mercury lamp (see Figure 4-21). In addition to mercury, argon, neon, and krypton, the arc tube contains halide salts (iodides) of metals. (Primary additives are mercury, sodium, and scandium iodides; others are thallium, indium, and cesium iodides.) These additives will add missing colors to the typical mercury lines, that is, red, orange, and yellow. The color of the metal halide is balanced across the spectrum. By improving the color without the need for a phosphor coating, the lamp approaches a point source that will result in better potential for optical control. For horizontal burning position, the arc tube is arced to make the temperature inside the arc tube more uniform (see Figure 4-21).

Starting time = 9 min (80%)
Testrike = 10 to 15 min (80%)

Enclosure

The outer enclosure (bulb) serves only two functions:

1. Ultraviolet filter.
2. Constant environment for the arc tube (maintains constant temperature and prevents air motion).

A phosphor coating is not necessary for good color rendition and should be avoided because of its negative effect on optical control; that is, the lamp no longer approaches a point source.

Electrical Connection

The metal halide lamp uses a mogul screw base for all wattages. Those lamps rated for horizontal (HOR) burning position that contain the curved arc tube (scc Figure 4-21) have a positioning pin on the base. A special positioning socket is available that assures proper positioning of the curved arc tube when the lamp engages the socket properly. The curved arc tube should always be positioned with the curve straight up in a vertical plane.

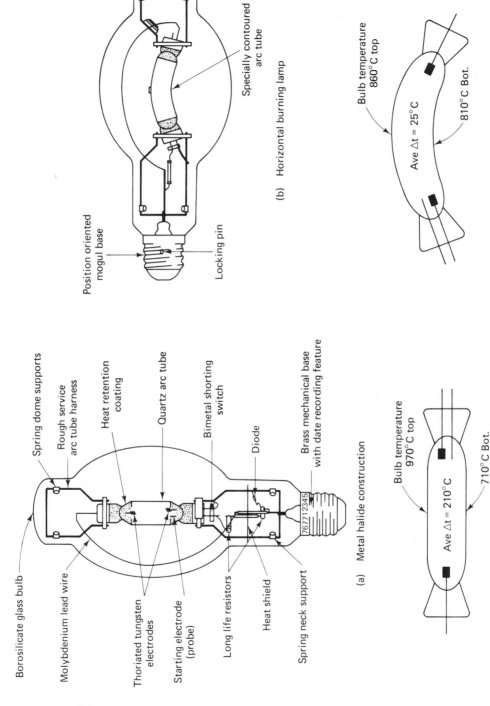

Position oriented
mogul base

Specially contoured
arc tube

Locking pin

(b) Horizontal burning lamp

Bulb temperature
860°C top

Ave ∆t = 25°C

810°C Bot.

Spring dome supports

Rough service
arc tube harness

Heat retention
coating

Quartz arc tube

Bimetal shorting
switch

Diode

Brass mechanical base
with date recording feature

Borosilicate glass bulb

Molybdenium lead wire

Thoriated tungsten
electrodes

Starting electrode
(probe)

Long life resistors

Heat shield

Spring neck support

767712345

(a) Metal halide construction

Bulb temperature
970°C top

Ave ∆t = 210°C

710°C Bot.

(c) Bulb wall and internal temperature variations for horizontal operation of a straight and curved arc tube.

Figure 4-21. Metal Halide Bulb Wall Temperature and Internal Temperature Variations (Courtesy GTE Sylvania, Incorporated)

Color Characteristics

The metal halide lamp produces energy at all wavelengths across the visible spectrum. That is, its SPD is well balanced, which means that the lamp produces good color rendition without the need for a phosphor coating (see Figure 4-22). The color appearance is a function of the quality control of the additives in the arc tube. Color consistency from one lamp to another is a function of the ballast, supply voltage, and lamp age. Where good color consistency (or match) between lamps is an important design consideration, the lamps should be replaced in groups rather than on a spot burn-out basis because of the change in color with life.

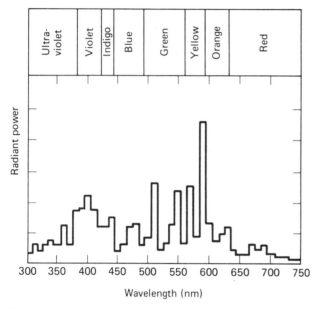

Figure 4-22. Spectral Power Distribution for Metal Halide Lamp (Courtesy of GTE Sylvania, Incorporated)

Lamp Designation

Lamp designations for metal halide lamps have not been standardized. The engineer must use caution in specifying lamps with nonstandard designations to make certain that a manufacturer is not being inadvertantly left out of a bid. The M or MH letter designation should be used to identify a metal halide lamp.

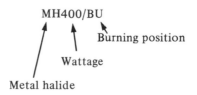

99

Metal halide lamps are especially sensitive to burning position. Manufacturers' data should be consulted for burning position requirements.

The bulb is designated by a letter, number combination. Metal halide lamps are available in BT and E shapes (see Figure 4-16). The number represents the maximum diameter of the outer bulb in eighths of an inch.

$$\text{Diameter} = \frac{37''}{8} = 4\frac{5''}{8}$$

BT-37

Performance Characteristics

LUMEN DEPRECIATION.   The lumen depreciation curve for a metal halide lamp is substantially better than for a mercury vapor lamp. The lumen output at the end of life for the higher-wattage lamp is 75 percent for a metal halide lamp (see Figure 4-23).

LIFE.   Life varies as a function of lamp wattage and the length of time that the lamp has been on the market. For example, the MH175/HOR was commercially available in 1972. The standard practice in the lamp industry is to bring all new lamps on the market with a rating of 7500 h. As data are developed on mortality and life, which requires long-term testing, lamp life is expected to increase to at least 15,000 h. Current lamp catalogs from *all* manufacturers should be consulted for life ratings of lamps.

ENERGY DISTRIBUTION.   The energy distribution for a metal halide lamp is shown in Figure 4-24.

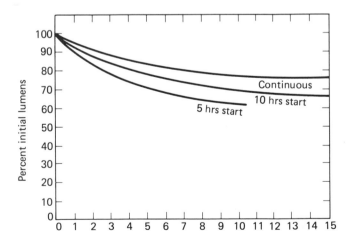

Figure 4-23. Metal Halide Lumen Depreciation (Courtesy of GTE Sylvania, Incorporated)

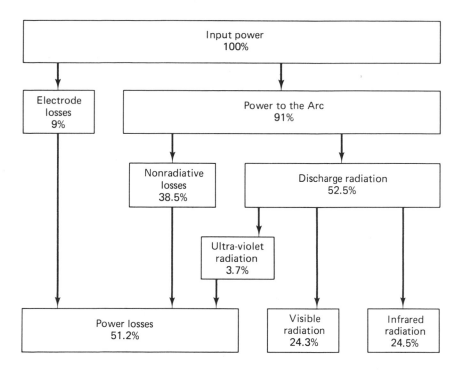

Figure 4-24. Energy Distribution for Metal Halide Lamp
100 lm/watt (40W) @ 24.3% visible radiation

LAMP EFFICACIES.   Lamp efficacies vary with burning position and lamp wattage. The higher the wattage, the higher the efficacy.

$$175 \text{ W: } 80 \text{ to } 85.7 \text{ lm/W}$$
$$250 \text{ W: } 82 \text{ lm/W}$$
$$400 \text{ W: } 85 \text{ to } 100 \text{ lm/W}$$
$$1000 \text{ W: } 100 \text{ to } 115 \text{ lm/W}$$
$$1500 \text{ W: } 96.7 \text{ to } 103.3 \text{ lm/W}$$

Note: Ranges of values are due to variations among manufacturers.

ENERGY-SAVING DEVICES.   Metal halide dimming is a recent development. The 400-W lamp can be dimmed (5 min) to 47 percent of its total input power, which results in a 22 percent reduction in lumens. The 1000-W metal halide can be dimmed (15 min) to 36 percent of its total input power, or 14 percent of its lumen output. As additional technological development occurs, the cost of dimming should decrease and the range should increase.

High-Pressure Sodium

Light-Producing Element

The light-producing element is an arc tube. The arc tube is small in diameter to maintain a high operating temperature. Because of the small diameter, there is no starting electrode inside the arc tube. Sodium, operating at a slight pressure and high temperature, will have a corrosive effect on ordinary glass or quartz. Therefore, the arc tube is made of a ceramic material (polycrystalline, translucent alumina material). The arc tube contains xenon, an amalgam of mercury, and sodium operated at a pressure of 200 mm of Hg.

$$\text{starting time} = 3 \text{ min } (80\%)$$
$$\text{restrike} = 1 \text{ min } (80\%)$$

Enclosure (Bulb)

The outer enclosure helps to maintain the arc tube in a constant-temperature environment and shields the arc tube from air motion.

Electrical Connection

The electrical connection is a mogul screw base. The lamp requires a 2500 to 5000 V pulse of energy to start the lamp. This is accomplished by a small electronic starting device that provides the high voltage pulse to break down the resistance and start the lamp.

Color Characteristics

The high-pressure sodium lamp produces energy at all wavelengths (Figure 4-25). However, the major portion of energy is concentrated in the yellow-orange part of the spectrum. The color characteristics of the lamp turn red objects orange and will darken the color appearance of blue and green objects. Increasing arc tube pressure seems to improve the color appearance of reds, blues, and greens. Color consistency from one lamp to another lamp is better than with metal halide lamps. However, color shifts can occur due to voltage variations and differences in ballasts.

Lamp Designation

High-pressure sodium lamp designations have not been standardized for the lamp industry. The engineer must use caution not to specify or use trade names that close acceptable lamps out of a bid. High-pressure sodium lamps are available in E, BT, and T shapes (see Figure 4-16). A letter, number combination is used to designate the bulb configuration.

Performance Characteristics

LUMEN DEPRECIATION.   The lumen depreciation curve for the high-pressure sodium lamp is the best of all high-intensity discharge (HID) lamps. The lumen output at the end of life for the higher-wattage lamps is 80 percent (see Figure 4-26).

LIFE.   Life varies as a function of wattage, ballast circuit, and manufacturer. The range is from 15,000 to 24,000 h for the more common higher-wattage lamps.

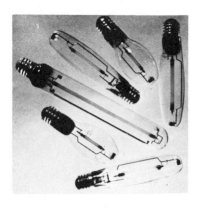

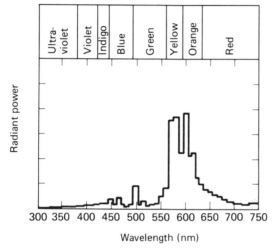

Figure 4-25. Spectral Power Distribution for High-Pressure Sodium Lamp (Courtesy of General Electric Company)

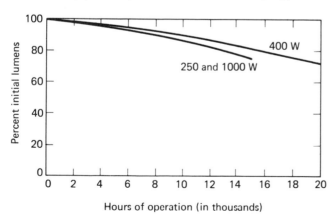

Figure 4-26. High-Pressure Sodium Lamp Lumen Depreciation (Courtesy of General Electric Company)

ENERGY DISTRIBUTION.   The energy distribution for the high-pressure sodium lamp is shown in Figure 4-27.

LAMP EFFICACIES.   High-pressure sodium lamp efficacies vary as a function of the burning position and the lamp wattage.

70 W: 77 to 82.9 lm/W
100 W: 88 to 95 lm/W
150 W: 100 to 106.7 lm/W
250 W: 102 to 120 lm/W
400 W: 118.8 to 125 lm/W
1000 W: 140 lm/W

The high-pressure sodium lamp is also available in wattages that can be operated on a mercury ballast. The wattages available are 150, 215, 310, 360 W. Manufacturers' data should be consulted for proper mercury ballast matching with the lamp.

ENERGY-SAVING DEVICES.   Dimming of some wattages of high-pressure sodium lamps is possible. The 1000-W lamp can be dimmed to 38 percent of its total input power in approximately 15 min, with a reduction in light output to 20 percent of rated lumens.

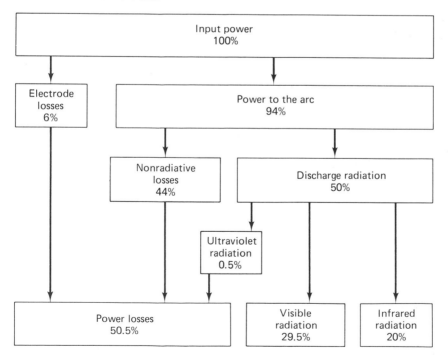

Figure 4-27. Energy Distribution for the High-Pressure Sodium Lamp 125 lm/watt @ 29.5% visible radiation.

BALLASTS

Ballasts are required on all gaseous discharge lamps. A ballast is referred to as a zero- or negative-resistance device.

Zero-Resistance Element

Since $I = E/R$ as $R \longrightarrow \infty$.

Ballast Function

The function of the ballast is to:

1. Provide starting voltage (voltage kick).
2. Limit current (as resistance $R \longrightarrow 0$).
3. Provide a power-factor correction (inductive reactance loading).

Fluorescent Ballasts

The American National Standards Institute (ANSI) sets the requirements to meet good lamp performance, such as light output, life, and reliability. Certified Ballast Manufacturers (CBM) is an organization that was established to assure the user that the ballast meets or exceeds the minimum ANSI standards. The rapid-start ballast provides a continuous heating voltage to the cathodes. The ballast supplies 4 V to the input sockets. The most common rapid-start ballast operates the lamps in series. (Supply leads from ballast to lamp are red, and energy returns to the ballast from the lamp via the blue leads.) A typical fluorescent ballast will show a 1 percent change in lumen output for each 1 percent of change in rated voltage. A metal strip is required to be within 1 in. of the lamp. The metal strip acts as a capacitor that causes an effective decrease in the lamp length, which results in lower starting voltage requirements.

Typical Ballast Circuits

The primary elements of a ballast can be seen in Figure 4-28, which shows the circuits for the preheat, instant-start (slimline), and rapid-start lamps. The ballast components are:

1. Core and coil: a reactive coil provides the voltage kick and is current limiting.
2. Autotransformer, which boosts starting voltage.
3. Capacitor, which provides a power-factor correction (phase).

Rapid-Start Ballast

Starting is a two-step process:

1. Conditioning the lamp: a continuous current preheats the cathode, causing an ionization of the argon or argon–neon. This preionization decreases the resistance of the lamp, allowing the second step.

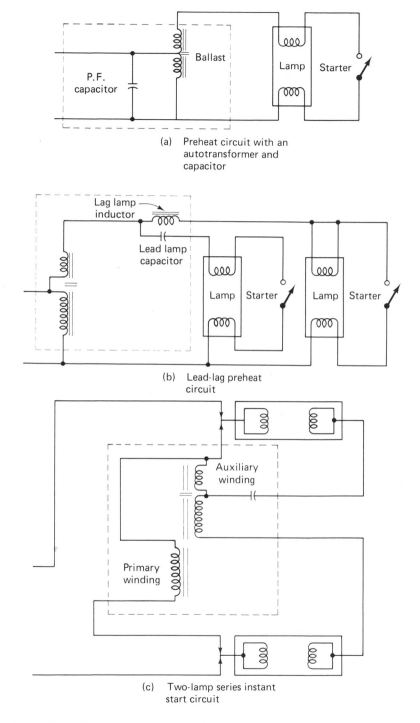

(a) Preheat circuit with an autotransformer and capacitor

(b) Lead-lag preheat circuit

(c) Two-lamp series instant start circuit

Figure 4-28. Typical Fluorescent Ballast Circuits (Courtesy of General Electric Company)

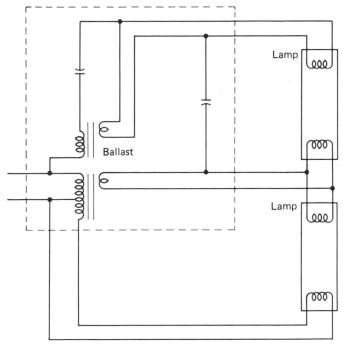

(d) Two-lamp series lead
     rapid start circuit

Figure 4-28 *(cont.)*

2. Striking the arc: decreased resistance means that a low voltage is required
   to strike the arc.

A trigger-start ballast is similar to the rapid-start ballast in that no starter is
required and a continuous current is supplied to the cathodes. It is designed for
use with preheat lamps of 20 W and less. It is available in a high- and low-power-
factor ballast.

Class P Ballast

The National Electric Code (NEC) states that "fluorescent fixtures for indoor installations shall incorporate ballast protection." Thermal protected
ballasts that complies with Underwriter Laboratories' requirements are marked
class P. The main requirements of class P ballast are as follows:

1. That the ballast not dangerously overheat as a result of either internal faults
   or external abnormal conditions.
2. That the calibration of the protective device remain unchanged under the
   anticipated conditions of use and operation.
3. That the device perform safely when called upon to operate.

107

With class P protection, any overheating problem caused by misapplication becomes extremely critical. The variables that affect the ballast operating temperature are as follows:

1. Supply voltage: case temperature rises 0.8 °C for every 1 V above the rated supply voltage (at 120 V).
2. Ambient temperature: a 1 °C rise in ambient temperature can raise the case temperature 0.9 °C.
3. Luminaire mounting: surface mounted luminaire can have temperatures as much as 10 °C higher depending on ceiling material and insulation.
4. Luminaire design: variations in number of lamps, enclosure, size, material, lenses, and ballast location can account for a 20 °C difference.
5. Ballast and lamp: manufacturing variations can affect ballast temperature by ±5 °C.

Under normal operating conditions in a 25 °C ambient, class P ballasts must meet the same requirements and limitations as all listed ballasts. These are the following:

1. 90 °C maximum case temperature.
2. 70 °C maximum capacitor temperature.
3. 65 °C maximum coil temperature rise.

Under fault conditions the continuous case temperature on class P ballasts is limited to 110 °C. A maximum of 110 °C case temperature under fault and a test case temperature of 90 °C means an allowable case temperature increase of only 20 °C. Thus, the engineer must know the environmental condition at each installation and specify a luminaire that will meet the design requirements.

Low-Pressure Sodium Ballasts

The conventional low-pressure sodium ballast is an autoleakage transformer maintaining nearly constant current, with an increase in lamp voltage throughout the lamp's life resulting in an increase in power consumption with age (see Figure 4-29). Open-circuit voltage for ignition and restrike is from 390 to 575 V. It is this high open-circuit voltage that accounts for the high ballast losses.

High-Pressure Gaseous Discharge (or HID) Ballasts

There are no standards for HID ballasts. Because of this, manufacturers are not held to any regulations, which results in a large variation in lamp performance. An important consideration in the selection of HID ballasts is the lamp cur-

Nominal watts	Lumens	Input watts (100 hours)	System* efficacy (100 hours)	Input watts (18,000)	System* efficacy (18,000)
35	4,650	60	77.5	68	68.4
55	7,700	80	96.3	89	86.5
90	12,500	125	100.0	157	79.6
135	21,500	178	120.8	226	95.1
180	33,000 (34,000)	220	150.0	285	119.3

*System efficacy is lamp plus ballast — ballast is a High Power Factor, High Reactance Type.

Nominal watts	Supply voltage	Starting voltage to lamp
35	120/240	390
55	120/240	410
90	120, 240, 277	420
135	120, 240, 277, 480	575
180	120, 240, 277, 480	575

Conventional 180 watt LPS ballast characteristics

Lamp efficacy (100 hr)	183.3 lumens/watt
Lamp efficacy (18,000 hr)	188.9 lumens/watt
Ballast watts (100 hr)	40
Ballast watts (18,000 hr)	105
Lumens (100 hr)	33,000
Lumens (18,000 hr)	34,000
Circuit efficacy (100 hr)	150 lumens/watt
Circuit efficacy (18,000 hr)	119.3 lumens/watt

Figure 4-29. Low-Pressure Sodium Lamp (Courtesy of General Electric Company)

rent crest factor. The lamp current crest factor is a ratio of the peak value of the wave to its effective (or rms) value. The maximum allowable range for the crest factor should be from 1.45 for a good reactor ballast to an upper limit of 2.0. The higher the crest factor, the greater the lumen depreciation of a lamp during its life. For best lamp performance, the preferred range should be from 1.6 to 1.8. Lamp manufacturers will provide the crest factor value necessary to produce the published lamp performance data. Most lamp-manufacturer data are based on a standard reference ballast (reactor type) with a crest factor of 1.4 to 1.5.

Mercury Vapor Ballasts

In general, if line voltage will vary more than ±5 percent, a regulating-type ballast should be used. If regulation is good, a reactor or autotransformer ballast with good crest factor should be used (Figure 4-30).

A reactor (inductive) has the following characteristics:

1. Nonregulating: input voltage is controlled over a narrow range; maximum allowable variation of input voltage is ±5 percent.
2. Available in high or low power factor.
3. Higher than normal starting current; requires heavier wire and circuit protection; primary starting current is 150 percent of operating current.
4. More efficient than regulating-type ballast.
5. Simple, light, quiet, least expensive.
6. Current crest factor of 1.45 to 1.55.
7. Minimum line voltages: 240 and 277 for all lamps except the 1000- and 1500-W lamps, which require 480-V supply.

An autotransformer (high-reactance, lag) has the following characteristics:

1. Nonregulating: input voltage variation of ±5 percent.
2. High or low power factor.
3. More efficient and quieter.
4. Current crest factor of 1.45 to 1.55.
5. Line voltage of 480 V for 1000- and 1500-W lamps, and 240 and 277 V for all others.
6. Higher cost than reactor.
7. Used where line voltage must be stepped up to meet the starting needs of the lamp.

A premium constant-wattage (CW, regulator, isolated regulator) has the following characteristics:

1. Regulation: lamp operation for large variations in line voltage of ±10 percent.

Mercury Lamp Ballasts

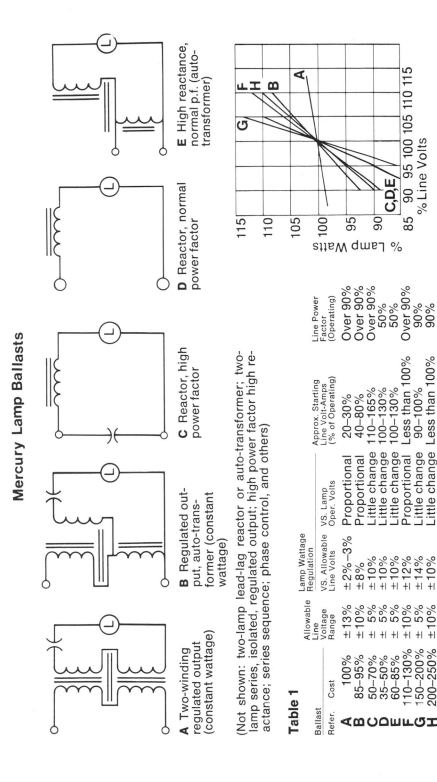

A Two-winding regulated output (constant wattage)

B Regulated output, auto-transformer (constant wattage)

C Reactor, high power factor

D Reactor, normal power factor

E High reactance, normal p.f. (auto-transformer)

(Not shown: two-lamp lead-lag reactor or auto-transformer; two-lamp series, isolated, regulated output; high power factor high reactance; series sequence; phase control, and others)

Table 1

Ballast Refer.	Cost	Allowable Line Voltage Range	Lamp Wattage Regulation		Approx. Starting Line Volt-Amps (% of Operating)	Line Power Factor (Operating)
			VS. Allowable Line Volts	VS. Lamp Oper. Volts		
A	100%	±13%	±2%–3%	Proportional	20–30%	Over 90%
B	85–95%	±10%	±8%	Proportional	40–80%	Over 90%
C	50–70%	±5%	±10%	Little change	110–165%	Over 90%
D	35–50%	±5%	±10%	Little change	100–130%	50%
E	60–85%	±5%	±10%	Little change	100–130%	50%
F	110–130%	±10%	±12%	Proportional	Less than 100%	Over 90%
G	150–200%	±5%	±14%	Little change	90–100%	90%
H	200–250%	±10%	±10%	Little change	Less than 100%	90%

Figure 4-30. Mercury-Vapor Circuit (Courtesy of Westinghouse Lamp Divison)

2. Lamp isolated from the line.

3. Most expensive form of ballast.

4. Current crest factor of 1.8 to 2.0, which is the poorest of all ballasts.

5. Can use all standard input line voltages.

6. A ±5 percent variation in lamp watts for line voltage range.

A constant-wattage autotransformer (CWA, autoregulator) has the following characteristics:

1. Most common compromise between regulation and cost.

2. Regulation: line voltage variation of ±10 percent allowable.

3. Smaller and lighter than premium constant wattage but larger and heavier than reactor.

4. Current crest factor of 1.6 to 1.8.

5. Noise and losses lower than premium constant wattage but worse than reactor.

Metal-Halide Ballasts

Metal halide lamps will not operate on a reactor ballast. Three types of ballast can be used on metal halide lamps.

1. Regulating

2. Autoregulated

3. Lead-peak autotransformer

    a. Similar to CWA, except regulation not as good.

    b. Regulation: input voltage of ±10 percent.

    c. Current crest factor of 1.6 to 1.8.

    d. Higher cost than mercury ballasts.

High-Pressure Sodium Ballast

The important characteristics associated with the high-pressure sodium ballast are as follows:

1. Larger and more complex; requires an auxiliary starter.

2. High voltage to the lamp.

3. Voltage varies over the life of the lamp. The operating characteristics are described in terms of the **trapezoidal** limits.

4. Current crest factor should not exceed 1.8.

Two types of ballasts are available for operating high-pressure sodium lamps: (1) regulating HPS ballast, +7.5 to 10 percent for 1000- and 400-W lamps; +10 percent for 250-W lamps, which operate on all standard voltages; (2) reactor HPS ballast, ±5 percent input voltage; minimum of 240 V on all lamps, except 480 V for 1000-W lamps.

Starting Voltages

Starting voltage or open-circuit voltage is the voltage required to initiate the electron stream between the electrodes. Open-circuit voltage is a function of lamp wattage. Starting voltages are as follows for the sources listed:

1. Low-pressure sodium (SOX 180): 575 V.
2. Mercury vapor: 225 to 450 V.
3. Metal halide: 245 to 450 V.
4. High-pressure sodium: 2500 to 5000 V.

Starting and Restrike Times (Summary)

Light source	Starting 80% (min)	Restrike 80%
Low-pressure sodium (180 W)	9–15	30 s
Mercury vapor	7–10	7 min
Metal halide	5	10–15 min
High-pressure sodium	3	1 min

LIGHT-SOURCE COMPARISONS

Tables 4-1 through 4-3 are provided to allow comparison of the seven light sources described in this chapter. Tables 4-1 and 4-2 compare light sources in terms of the lamp only (no ballast losses), using constant wattage (400 W) and constant lumens (approximately 30,000 lm), respectively. Table 4-3 compares the light sources with the combined effect of lamp plus ballast loss. This gives a more meaningful picture of lamp performance. However, the results can still lead to erroneous conclusions if the luminaire efficiency (distribution) is not included in the analysis (see Chaps. 7 and 9). The total system (lamp + ballast + luminaire) must be analyzed completely before making design decisions on equipment selection.

TABLE 4-1

Source Comparison Based on Equal Watts (400 W)

Lamp	Design	Quantity	Watts (total)	Lumens (each)	Source efficacy (each)	BTU per hour (total)	Life in hours (each)	Total cost (each)[b]
Incandescent	100A19	4	400	1,740	17.4	1364	750	(0.50)[a] 2.00
Tungsten–halogen	Q400T4/CL	1	400	7,500	18.8	1364	2,000	16.25
Fluorescent	F40CW	10	400	3,150	78.9	1364	20,000	(1.67) 16.70
Low-pressure sodium	SOX135	3	405	21,500	159.3	1381	18,000	(40.00) 120.00
Mercury vapor	H33GL-400/DX	1	400	22,500	56.3	1364	24,000[a]	15.50
Metal halide	M400/BU-HOR	1	400	34,000	85.0	1364	15,000	34.50
High-pressure sodium	LU400/BU	1	400	50,000	125.0	1364	20,000	60.00

[a]Rated life is 24,000 h; usable life is 16,000 to 18,000 h.

[b]Values in parenthesis are costs per individual lamp.

114

TABLE 4-2

Source Comparison Based on Equal Lumens (30,000 lm)

Lamp	Design	Quantity	Total Lumens (each) [b]	Watts (each)	Source efficacy (each)	BTU per hour (total)	Life in hours (each)	Total cost (each) [b]
Incandescent	100A19	17	29,580 (1,740)	100	17.4	5797	750	(0.50) 8.50
Tungsten–halogen	Q400T4/CL	4	30,000 (7,500)	400	18.8	5456	2,000	(16.25) 65.00
Fluorescent	F40CW	10	31,500 (3,150)	40	78.9	1364	20,000	(1.67) 16.70
Low-pressure sodium	SOX180	1	33,000 26,000	180	183.3	614	18,000	60.00
Mercury vapor	H37KC-250/DX	2	(13,000)	250	52.0	1705	24,000[a]	(18.75) 37.50
Metal halide	M400/BU-HOR	1	34,000	400	85.0	1344	15,000	34.50
High-pressure sodium	LU250/BU/S	1	30,000	250	120.0	853	15,000	64.00

[a]Rated life is 24,000 h; usable life is 16,000 to 18,000 h.

[b]Values in parenthesis are for individual lamps.

TABLE 4-3

Light Source Summary

	Designation	Watts	Lumens	Source efficacy	System efficacy	Life
Incandescent						
Standard	100A19	100	1,750	17.5	17.5	750
Tungsten–halogen	Q250CL/MC	250	4,850	19.4	19.4	2,000
Gaseous Discharge						
Low pressure						
Fluorescent	F40CW	40	3,150	78.8	68.5[a]	20,000
Low-pressure sodium	SOX180	180	33,000	183.3	{ 150.0 / 119.3[b]	18,000
High-pressure						
Mercury vapor	H33GL-400/DX	400	22,500	56.3	49.2[c]	24,000
Metal halide	MS400/HOR	400	40,000	100.0	85.0[d]	15,000
High-pressure sodium	LU400/BU	400	50,000	125.0	100.2[e]	24,000

[a]HPF-RS two-lamp ballast, 92 W.

[b]HPF, high reactance, 220 W at 33000 lm (100 h) to 285 W at 34000 lm (18,000 h).

[c]Regulating, 457 W.

[d]Autostabilized, 471 W.

[e]Stabilized, 499 W.

references

1. Kaufman, John, editor, **IES Lighting Handbook,** 5th ed., Illuminating Engineering Society, New York, 1972.

2. Koedam, M., R. DeVann, and T. G. Verbeek, "Further Improvements of the LPS Lamp," **Lighting Design and Application,** Sept. 1975, pp. 39–45.

5

luminaires

A **luminaire** is the lighting equipment that houses the light source, electrical components, and light-control method. This chapter describes the various means of redirecting and controlling the light from a luminaire. The measurement of the light output and the calculations required to develop a report (photometric test report) that contains important design parameters will be discussed. The factors that affect in-place performance will be presented as well as cost considerations.

LIGHT CONTROL

The light sources discussed in the previous chapter provide us with a source of luminous flux. This flux is not necessarily directed where it will do the most good. For example, a bare incandescent lamp radiates in all directions. To control or redirect the light to where we want it, materials are used that have one or more of three basic properties: transparency, translucency, or opacity. Transparent materials will pass or transmit most of the light, but will refract it as it is transmitted. Opaque materials on the other hand will transmit no light; they reflect and absorb all the luminous energy.

When light strikes a material that has any one of the above properties it can be reflected, transmitted, refracted, absorbed, or polarized. One or more of these phenomena may occur simultaneously.

Reflection

The first method for controlling light, reflection, occurs in many forms: regular or specular reflection, spread reflection, diffuse reflection, mixed reflection, scattered reflection, and selective reflection (Figure 5-1).

1. **Regular or specular reflection** is the type that occurs in mirrors or highly polished metals. In such cases, the angle of reflection equals the angle of incidence.

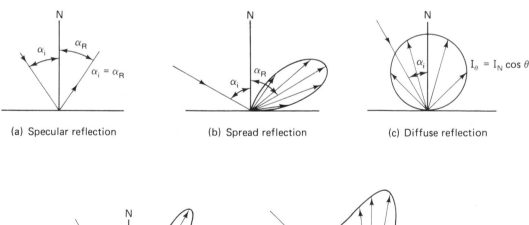

(a) Specular reflection   (b) Spread reflection   (c) Diffuse reflection

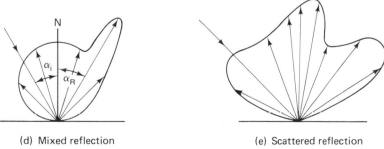

(d) Mixed reflection   (e) Scattered reflection

Figure 5-1. Reflection Patterns

2. **Spread reflection** is similar to specular reflection, except the light is scattered about the ray at the angle of reflection.

3. **Diffuse reflection** is exhibited by matte white paper or flat white paint. A perfect diffuser would obey **Lambert's cosine law,** which states that the intensity reflected at angle θ is equal to the intensity reflected normal to the surface times the cosine of the angle θ . The reflection pattern then is not a function of the angle of incidence. This is shown in Figure 5-1.

4. **Mixed reflection** is a combination of specular, spread, and diffuse reflection. Mixed reflection probably best describes real materials.

5. **Scattered reflection** is that reflection which cannot be associated with either Lambert's cosine law or the law of regular reflection. Figure 5-1 shows a possible reflection pattern.

6. **Selective reflection** gives objects their color by the phenomena called **selective absorption** (Chap. 2). Certain wavelengths of incident radiation are absorbed by the surface while others are reflected. The selection of which wavelength to reradiate determines the color of the object. Selective reflection can have any of the reflection patterns previously mentioned.

Control by Reflection

Generally, specular reflection is used to control light. There are four specific contours used in reflective light control: the **flat plane,** the **circle,** the **parabola,** and the **ellipse.** These basic shapes can be configured in numerous ways to direct light. For a flat plane, all incident rays are reflected at an angle equal to the angle of incidence. If the incident rays are parallel, the reflected rays will be parallel. For a circular reflector with the source at the center, all reflected rays return through the center. If the source is placed ahead of the center, the reflected rays will diverge as in Figure 5-2. Probably the best known reflector shape is the parabola. If the light source is placed at the focal point of a parabolic reflector, all reflected rays will be parallel (Figure 5-2). This is the principle used in a searchlight to obtain a beam of parallel light rays. The final shape is the ellipse. With an ellipsoidal reflector, a source at the first focal point of the ellipse causes the reflected rays to pass through the second focal point. This principle is used in the pinhole spot from which a wide pool of light is desired from a luminaire with a small ceiling opening or aperture.

Transmission

Transmission of light through specific materials occurs in patterns similar to light reflected off materials. The types of transmission are regular, spread, diffuse, mixed, scattered, and selective. Each type produces a distribution corresponding to the reflection patterns described above (see Figure 5-1), except on the side of the surface opposite the incidented ray.

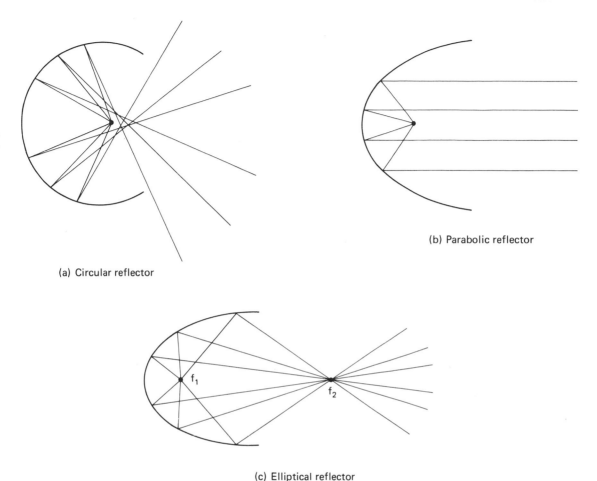

(a) Circular reflector

(b) Parabolic reflector

(c) Elliptical reflector

Figure 5-2. Reflector Shapes

Refraction

Snell discovered that as light travels from one media to another its speed changes slightly. As a consequence, the light rays are bent. This is known as **refraction,** and follows Snell's law,

$$n_1 \sin \alpha_1 = n_2 \sin \alpha_2$$

where n is the index of refraction and α is the angle between the normal to the surface and the ray (Figure 5-3). As the light ray travels from a less dense (air) medium through a more dense (glass) medium, the light ray is bent toward the normal to the surface. As the light ray leaves the dense medium and reenters the

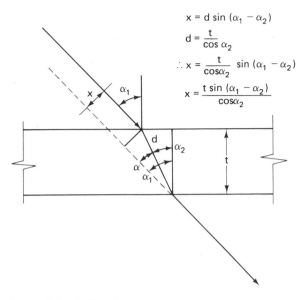

Figure 5-3. Refraction

less dense (air) medium, the ray is bent away from the normal. The emerging ray of light will be parallel to the entering ray of light but displaced a distance X. The displacement (X) can be calculated according to the formula derived in the figure. The phenomenon of refraction [2,3] is used in flat plates, lenses, and prisms to bend light.

Absorption

All materials, whether they transmit or reflect light, absorb some of the energy. Absorption can be either **selective** or **nonselective.** Selective absorption, as was stated earlier, is what causes objects to have color. Nonselective absorption does not affect the color appearance of the material. All wavelengths are absorbed equally in nonselective absorption, causing a change in the luminance of the material. An example of nonselective absorption in a transmitting material is a neutral density filter, which is used to change the intensity of a light ray without changing its wavelength composition.

Polarization

Light is composed of waves oscillating in an infinite number of planes perpendicular to the direction of propagation. Certain materials reflect or transmit only a portion of those waves. This light is said to be **polarized.** Figure 5-4 is a schematic representation of unpolarized light. When unpolarized light passes through a polarizing material, such as that found in many sunglasses, only certain waves are transmitted. Those that are blocked depend on the orientation of the polarizer. Most materials exhibit polarizing characteristics. Ordinary glass

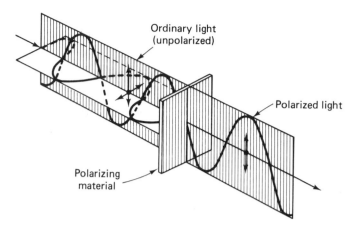

Figure 5-4. Polarization of Light

will block certain waves reflected from the surface. Maximum blocking occurs at Brewster's angle, which is 57° for ordinary glass (Figure 5-5).

LUMINAIRE PHOTOMETRICS

Once a light source has been chosen and the hardware designed to direct the light, luminaire performance must be ascertained. To determine this a **luminous intensity distribution** (more commonly known as a **candlepower distribution**) **test** is performed. This measurement and the maximum luminance are the two basic measurements used to generate the data from which a luminaire **photometrics test report** is produced. A photometric test report includes such information as (1) the luminous intensity distribution, (2) luminaire efficiency, (3) CIE–IES classifica-

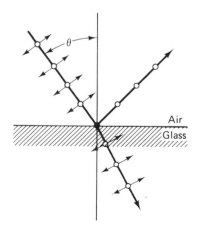
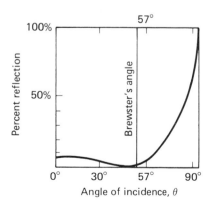

Figure 5-5. Brewster's Angle

tion, (4) coefficients of utilization, (5) maximum spacing, (6) maximum luminance, (7) average luminance, and (8) the maximum to average luminaire luminance.

Luminance Intensity

The most important set of data in artificial lighting design is the luminous intensity data. The luminous intensity distribution represents the intensity of luminous energy from a luminaire in a particular direction. Figure 5-6 is a plot of a typical luminous intensity distribution. This curve connects the heads of all the intensity vectors to form a distribution solid. The radial intensity vectors are plotted against a vertical angle measured from nadir on a polar coordinate system for a particular vertical plane through the luminaire. The center of the plot is the center of the luminaire, the nadir is defined as straight down from that point. Ideally, the intensity at an infinite number of vertical angles and vertical planes should be measured. Since this is impractical, a limited number of vertical planes

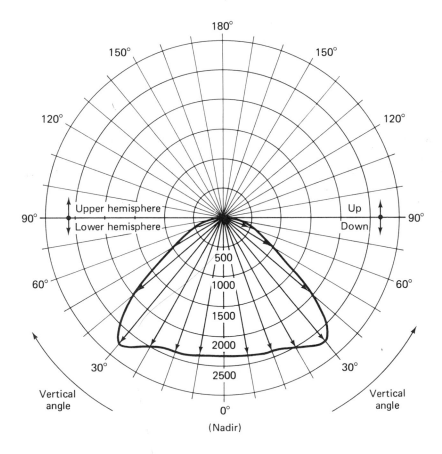

Figure 5-6.   Luminous Intensity (Candlepower) Distribution Curve.

and vertical angles are selected to define the luminous intensity distribution. The minimum number of vertical planes for an asymmetric distribution is three: 90° (perpendicular) to the lamp axis, 45°, and 0° (parallel) to the lamp axis. Two other planes should be measured, 22.5° and 67.5°, for greater accuracy. The minimum number of vertical angles should be in 10° steps. Preferably, vertical angles should be at 2½° steps. Figure 5-7 is a plan view of a 2- by 4-ft fluorescent luminaire showing the vertical measurement planes; the vertical angles or steps are indicated in Figure 5-6.

Average Luminous Intensity

The average luminous intensity is used to determine zonal lumens, luminaire efficiency, CIE–IES luminaire classification, coefficient of utilization, maximum spacing, and average luminaire luminance.

The luminaire is assumed to be placed in the center of a unit sphere. The unit sphere is divided into horizontal slices (Figure 5-8) or zones that are consistent with the way in which the luminous intensity data are recorded. The angles θ_1 and θ_2 define the width of a zone. The I_θ value in each vertical plane is measured halfway between θ_1 and θ_2. I_θ is referred to as the mid-zonal candlepower (luminous intensity). The mid-zonal candlepower values in each plane for a given zone are averaged together to give the average mid-zonal candlepower or average luminous intensity. For example, the equation to calculate the average luminous intensity in the zone between 40° and 50° of a luminaire with five-plane photometry would be

$$I_{ave\,55} = \frac{I_0 + 2I_{22.5} + 2I_{45} + 2I_{67.5} + I_{90}}{8}$$

Note in Figure 5-7 that there is only one 0° plane and one 90° plane, while there are two planes each at 22.5, 45, and 67.5°. Therefore, the denominator is 8 for

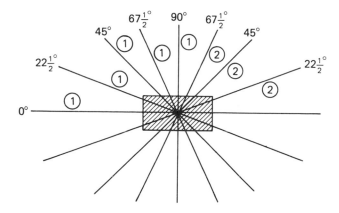

Figure 5-7. Vertical Planes Through Luminaire

five-plane photometry rather than 5. For three-plane photometry, the denominator will be 4, that is, one 0° plane, one 90° plane, and two 45° planes. Using the luminous intensity values given in Table 5-8 in the Data Section, the average luminous intensity values are calculated as follows:

$$I_{\text{ave}_5} = \frac{1388 + (2 \times 1390) + (2 \times 1393) + (2 \times 1385) + 1391}{8}$$

$$= 1389$$

$$I_{\text{ave}_{15}} = \frac{1336 + (2 \times 1344) + (2 \times 1353) + (2 \times 1363) + 1370}{8}$$

$$= 1353$$

The average luminous intensity is calculated in a similar fashion for the remaining mid-zonal angles (see Table 5-1).

TABLE 5-1

Average Luminous Intensity for Luminaire Table 5-8

Zone	Mid-zonal angle	Average candlepower	Zone	Mid-zonal angle	Average candlepower
0–10	5	1389	90–100	95	106
10–20	15	1353	100–110	105	116
20–30	25	1285	110–120	115	116
30–40	35	1159	120–130	125	115
40–50	45	937	130–140	135	96
50–60	55	615	140–150	145	75
60–70	65	365	150–160	155	63
70–80	75	221	160–170	165	49
80–90	85	135	170–180	175	39

Zonal Lumens

Zonal lumens are calculated to show the total amount and direction of light radiated from a luminaire. The light radiated from the luminaire is described in terms of the lumens produced on each zone or horizontal slice (Figure 5-8) of the unit sphere. The surface area of each horizontal strip can be calculated by

$$A_s = 2\pi (\cos \theta_2 - \cos \theta_1)$$

The area, A_s, of each horizontal strip is known as the zonal or lumen constant for the zone between θ_1 and θ_2. The zonal constants for 5 and 10° zones can be found in Table 5-9 in the data section.

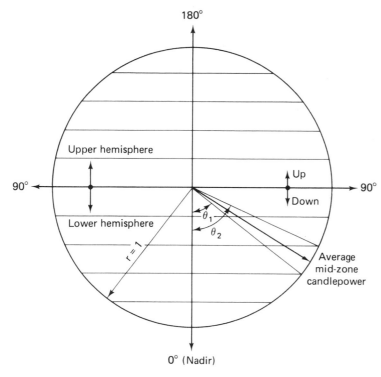

Figure 5-8. Unit Sphere for Calculating Zonal Constant

From the definition of luminous intensity and solid angle for a unit sphere $(r = 1)$,

$$I = \frac{d\phi}{d\omega}$$

$$\omega = \frac{A_s}{r^2} = \frac{A_s}{(1)^2} = A_s$$

we obtain the equation for calculating the luminous flux on a zone (zonal lumens) whose area is A_s on a unit sphere:

$$\phi = I \times A_s$$

Therefore, to calculate the luminous flux, ϕ, on a zone, we need the area of the zone, that is the zonal constant (ZC) (Table 5-9) and the average luminous intensity (Table 5-1). Zonal lumens (ZL) are calculated as follows:

$$
\begin{aligned}
\text{ZL} \quad &= \text{average luminous intensity} \times \text{ZC} \\
\text{ZL}_5 \quad &= 1389 \times 0.095 = 132 \\
\text{ZL}_{15} \quad &= 1353 \times 0.283 = 383
\end{aligned}
$$

The zonal lumens from Table 5-2 are summed for various key vertical angles and given in table 5-3.

TABLE 5-2

Zonal Lumens for Luminaire Table 5-8

Zone	Zonal constant	Zonal lumens	Zone	Zonal constant	Zonal lumens
5	0.095	132	95	1.091	116
15	0.283	383	105	1.058	123
25	0.463	595	115	0.993	115
35	0.628	728	125	0.897	103
45	0.774	725	135	0.774	74
55	0.897	552	145	0.628	47
65	0.993	362	155	0.463	29
75	1.058	234	165	0.283	14
85	1.091	146	175	0.095	4

TABLE 5-3

Lumen Summation

Vertical angles		
0–40		1838
40–90	(direct glare zone)	2019
		3857 (lower hemisphere)
90–180		625 (upper hemisphere)
0–180		4482 (total output)

Luminaire Efficiency

The luminaire efficiency is a ratio of the amount of light (lumens) that the luminaire puts out to that produced by the bare light source (see Table 5-4).

TABLE 5-4

Percentage of Total Lamp Lumens

$$\% \text{ TLL}_{0\text{-}40} = \frac{1838}{2 \times 3110} \times 100 = 29.5$$

$$\% \text{ TLL}_{40\text{-}90} = \frac{2019}{2 \times 3110} \times 100 = 32.5$$

$$\% \text{ TLL}_{90\text{-}180} = \frac{625}{2 \times 3110} \times 100 = 10.1$$

$$\% \text{ TLL}_{0\text{-}180} = \frac{4482}{2 \times 3110} \times 100 = 72.1$$

Luminaire efficiency $= \% \text{ TLL}_{0\text{-}180} = 72.1$

Luminaire efficiency is expressed as a single number and does not indicate where the luminous flux is directed.

$$\% \text{ Total lamp lumens} = \frac{\text{lumens out of luminaire}}{\text{bare lamp lumens}}$$

CIE–IES Luminaire Classification

The CIE–IES classification system describes a luminaire by the percentage of light in the upper and lower hemispheres of the luminous intensity distribution curve. The six categories (Figure 5-9) of the CIE–IES classification are (1) direct,

Classification	% Up light	% Down light	Typical candlepower distribution curve
Direct	0-10%	90-100%	
Semi-direct	10-30%	60-90%	
Direct-indirect *	40-60%	60-40%	
General diffuse	60-40%	40-60%	
Semi-indirect	60-90%	10-30%	
Indirect	90-100%	0-10%	

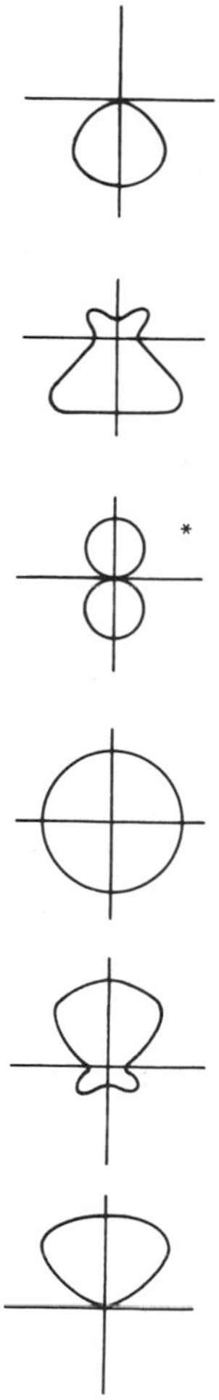

*IES classification only

Figure 5-9. CIE–IES Classifications

(2) semi–direct, (3) direct–indirect, (4) general diffuse, (5) semi–indirect, and (6) indirect. The general diffuse category has a horizontal component, whereas the direct–indirect (not recognized by the CIE) has very little horizontal component.

To determine the CIE–IES classification, the percentage of total output in the upper and lower hemisphere must be determined.

$$\% \text{ total output} = \frac{\text{lumens out in the hemisphere}}{\text{total lumens from 0 to 180}}$$

For the luminaire of Table 5-5, the CIE–IES luminaire classification is semi–direct.

TABLE 5-5

Percentage of Total Output in Upper and Lower Hemispheres

Lower hemisphere

$$\% \text{ TO}_{0\text{-}40} = \frac{1838}{4482} \times 100 = 41.0$$

$$\% \text{ TO}_{40\text{-}90} = \frac{2019}{4482} \times 100 = \underline{45.1}$$
$$\phantom{\% \text{ TO}_{40\text{-}90} = \frac{2019}{4482} \times 100 = } 86.1\% \text{ down}$$

Upper hemisphere

$$\% \text{ TO}_{90\text{-}180} = \frac{625}{4482} \times 100 = 13.9\% \text{ up}$$

Maximum and Average Luminance

Maximum and average luminaire luminances are provided by the manufacturer as part of a photometric test report. The average luminance is calculated for each of the glare angles (45, 55, 65, 75, and 85°) utilizing luminous intensity (candlepower) data and the projected area of the luminous surface area. Examples of the different types of luminaires and what to use for the projected area are shown in Figure 5-10.

The projected area across (cross axis viewing) and along must be calculated. Table 5-6 shows the calculated projected areas utilizing the specific formulas for the luminaire in Table 5-8 of the data section.

$$PA_{\text{across}} = L \times [(h \sin \theta) + (w \cos \theta)]$$

$$PA_{\text{along}} = L \times w \times \cos \theta \quad \text{(metal ends)}$$

To determine average luminance, the following equation is used:

$$\text{average luminance (fL)} = \frac{\text{luminous intensity at } \theta \text{ (candela)} \times 144\pi}{\text{project area at } \theta \text{ (square inches)}}$$

$$I_{\text{ave}} = \frac{I_\theta}{PA_\theta} \times 144\pi$$

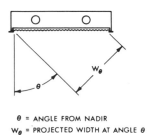

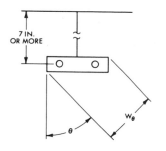

θ = ANGLE FROM NADIR
W_θ = PROJECTED WIDTH AT ANGLE θ

LUMINAIRE TYPE A

LUMINOUS SIDES AND BOTTOM 150 FL OR MORE.
IF SIDES ARE LESS THAN 150 FL, THEY ARE OMITTED
FROM W_θ.

LUMINAIRE TYPE B

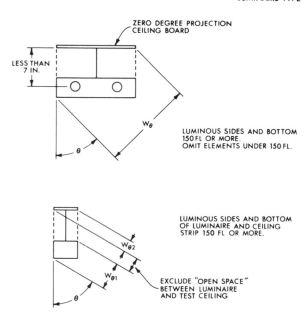

ZERO DEGREE PROJECTION
CEILING BOARD

LUMINOUS SIDES AND BOTTOM
150 FL OR MORE.
OMIT ELEMENTS UNDER 150 FL.

LUMINAIRE TYPE C

LUMINOUS SIDES AND BOTTOM
OF LUMINAIRE AND CEILING
STRIP 150 FL OR MORE.

EXCLUDE "OPEN SPACE"
BETWEEN LUMINAIRE
AND TEST CEILING

Figure 5-10. Projected Area (Courtesy of the Illuminating Engineering Society of North America)

TABLE 5-6

Projected Area for the Luminaire in Table 5-8

	Across (90°)			Along (0°)		
Angle	PW	L	PA	PL	W	PA
45	9.19	46.5	427.3	32.9	10	329
55	8.19	46.5	380.8	26.7	10	267
65	6.95	46.5	323.2	19.7	10	197
75	5.49	46.5	255.3	12.0	10	120
85	3.86	46.5	179.5	4.1	10	41

The constant, 144π, is used to convert the average luminance in candelas per square inch (cd/in.^{2}) to footlamberts (fL) (see Chap. 3).

Using the luminaire photometrics in Table 5-8 in the data section, the average luminaire luminance is calculated as follows and tabulated in Table 5-7:

$$\frac{\text{along } (0^\circ):}{L_{ave}45^\circ} = \frac{835}{329} \times 144\pi = 1148 \text{ fL}$$

$$\frac{\text{across } (90^\circ):}{L_{ave}45^\circ} = \frac{1018}{427.3} \times 144\pi = 1076 \text{ fL}$$

TABLE 5-7

Average Luminance, Maximum Luminance, and Maximum-to-Average Ratio

	Along (0°)					Across (90°)				
Angle	I_{ave}	PA	L_{ave}	Max L	M/A	I_{ave}	PA	L_{ave}	Max L	M/A
45	835	329	1148	2483	2.2	1018	427.3	1076	2721	2.5
55	537	267	909	1939	2.1	698	380.8	827	2211	2.7
65	295	197	678	1089	1.6	450	323.2	628	1701	2.7
75	142	120	533	850	1.6	317	255.3	560	1429	2.6
85	35	41	384	510	1.3	237	179.5	597	1463	2.5

The maximum luminaire luminance (Table 5-8, data section) is a measured quantity. A photometer with a circular aperture of 1 in.^{2} area is used to search the surface of the lens for the maximum luminance at the glare angles. The ratio of the maximum-to-average luminaire luminance is computed and used as an indicator of visual comfort (see Chap. 7). The maximum-to-average luminaire luminances are given in Table 5-7.

Coefficients of Utilization and Maximum Spacing Ratios

Coefficients of utilization and maximum spacing ratios are calculated from the luminous intensity distribution data. These two parameters are used in the lumen method of design, and are discussed in more detail in Chap. 7.

Standard Photometric Test Report

The previously calculated data are combined and given on a standard photometric test report. An example of a complete photometer test report for the luminaire analyzed is given in Figure 5-11 in the data section. Additional photometric data is available in reference 5 for working additional examples.

LUMINAIRE SELECTION PARAMETERS

The section on luminaire photometrics was concerned with how the luminaire performs relative to its distribution of luminous energy. Equally important is how it performs mechanically. Such items as heat, noise, maintenance, and cost must be considered in the choice of a luminaire. The heat and noise produced by a luminaire are discussed in Chap. 6.

Maintenance

Luminaires must be periodically maintained to keep them at peak performance. Maintenance includes changing lamps and cleaning the surfaces that collect dirt. To facilitate lamp changing, the luminaire should be constructed so that maintenance personnel can easily and rapidly reach the lamps. A latch and hinge arrangement is best because it avoids handling of the lens. The construction of the luminaire should be examined in terms of its rigidity, strength, and durability. These factors will determine the long-term usefulness of the luminaires. The design of the luminaire will affect the infiltration and movement of dirt inside the luminaire, which results in dirt accumulation and light loss. Any type of luminaire must be periodically cleaned. The time between cleanings can be lengthened if the luminaire is tightly constructed and designed to exclude airborne dirt; however, surface dirt accumulation will still occur and require cleaning.

A good maintenance program plays a key role in energy conservation. Lighting calculations take light losses due to poor maintenance programs into consideration. The net result is an increase in the number of luminaires to compensate for light loss due to inadequate maintenance. An increase in the number of luminaires means an increase in the watts for lighting, which is counterproductive to energy conservation. Poor maintenance results in the loss of luminous energy, which is absorbed by elements in the environment. Good maintenance will reduce operating cost and conserve energy.

Cost

Good light control and sound construction are paramount for a lighting system to perform properly for many years. Like any item of quality, a luminaire that has the best operating and maintenance characteristics is usually the most expensive. Cost will be discussed in detail in Chap. 7.

Additional Selection Factors

Numerous additional factors, such as comfort, appearance, light source, and application, are involved in the selection of a luminaire. Multiply this by the number of manufacturers and one has many thousands of luminaires from which to choose. A system proposed by Helms[4] would computerize the factors involved, and selection of luminaires could be made based on the parameters input by the designer.

It is important to realize that slight changes in the construction of the reflector or a change in lens material will result in a change in the luminous intensity distribution. This means that the efficiency, average luminance, maximum luminance, and the coefficient of utilization will change, even though the outward appearance of the luminaire has not changed significantly.

references

1. Kaufman, John, editor, **IES Lighting Handbook,** 5th ed. Illuminating Engineering Society, New York, 1972.

2. Kingslake, R., **Applied Optics and Optical Engineering,** Academic Press, New York, 1965.

3. Williams, C. S., **Optics: A Short Course for Engineers and Scientists,** John Wiley & Sons, Inc. (Interscience Division), New York, 1972.

4. Helms, R. N., "Luminaire Selection and Design—An Alternative to Manufacturer's Catalogs," **Lighting Design and Application,** Aug. 1974, p. 26.

5. Helms, R. N., **Illuminating Engineering Workbook,** 2nd Edition, University of Colorado Publication Service, Boulder, Colorado, 1979.

data section

TABLE 5-8

Basic Test Data

Photometric data

Description:

1 x 4 Surface mounted clear
prismatic acrylic wraparound

Test No. 11417 Cat. No. BK240
Lamps: 2F40CW Volts 120
Lumens/Lamp: 3110 Bare Lamp: 2360 fL
Shielding: Parl 90° Norm. 90°

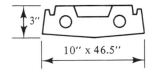

Luminous intensity (candlepower) candelas

Angle	0°	22½°	45°	67½°	90°
5	1388	1390	1393	1385	1391
15	1336	1344	1353	1363	1370
25	1234	1260	1289	1309	1331
35	1074	1119	1167	1200	1228
45	835	890	948	982	1018
55	537	573	615	657	696
65	295	325	357	405	450
75	142	174	212	269	317
85	35	83	134	186	237
95	10	50	100	163	214
105	23	71	115	163	210
115	33	78	120	154	191
125	41	83	120	148	176
135	43	67	93	124	153
145	49	64	81	86	88
155	46	54	66	71	74
165	39	44	50	54	57
175	37	38	39	41	42

Maximum luminance fL

Angle	0°	90°
45	2483	2721
55	1939	2211
65	1089	1701
75	850	1429
85	510	1463

S/MH = 1.4

135

TABLE 5-9

Zonal Constants: $2\pi(\cos\theta_1 - \cos\theta_2)$

5° Zones		
Zones		
Lower	Upper	Zonal constant
0–5	175–180	0.0239
5–10	170–175	0.0715
10–15	165–170	0.1186
15–20	160–165	0.1649
20–25	155–160	0.2097
25–30	150–155	0.2531
30–35	145–150	0.2946
35–40	140–145	0.3337
40–45	135–140	0.3703
45–50	130–135	0.4041
50–55	125–130	0.4349
55–60	120–125	0.4623
60–65	115–120	0.4862
65–70	110–115	0.5064
70–75	105–110	0.5228
75–80	100–105	0.5351
80–85	95–100	0.5434
85–90	90–95	0.5476

10° Zones		
Zones		
Lower	Upper	Zonal constant
0–10	170–180	0.095
10–20	160–170	0.283
20–30	150–160	0.463
30–40	140–150	0.628
40–50	130–140	0.774
50–60	120–130	0.897
60–70	110–120	0.993
70–80	100–110	1.058
80–90	90–100	1.091

PHOTOMETRIC DATA

SKETCH:

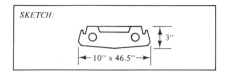

Description:

1 x 4 surface mounted clear
prismatic acrylic wraparound
w/metal ends

Test No. 11417   Cat. No. BK 240
Lamps   2F40 CW   Volts 120
Lumens/Lamp 3110   Test Dist. 20
Shielding Angle: N 90°   L. 90°
Bare Lamp 2360 fL   Date

LDD CATEGORY V

Brightness Data			
Angle	Maximum	Average	Ratio Max./Min.
Alonge (0°)			
45°	2483	1148	2.2
55°	1939	909	2.1
65°	1089	678	1.6
75°	850	533	1.6
85°	510	384	1.3
Across (90°)			
45°	2721	1076	2.5
55°	2211	827	2.7
65°	1701	628	2.7
75°	1429	560	2.6
85°	1463	597	2.5

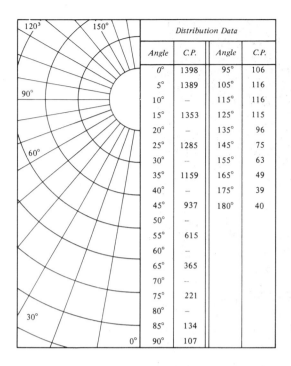

Distribution Data			
Angle	C.P.	Angle	C.P.
0°	1398	95°	106
5°	1389	105°	116
10°	–	115°	116
15°	1353	125°	115
20°	–	135°	96
25°	1285	145°	75
30°	–	155°	63
35°	1159	165°	49
40°	–	175°	39
45°	937	180°	40
50°	–		
55°	615		
60°	–		
65°	365		
70°	–		
75°	221		
80°	–		
85°	134		
90°	107		

OUTPUT DATA			
Zone Degrees	Lumens	Percent Total Lamp Lumens	Percent Total Distance
0–40°	1838	29.5	41.0
40–90°	2019	32.5	45.1
90–180°	625	10.1	13.9
0–180°	4482	72.1	100.0

COEFFICIENT OF UTILIZATION

Floor		ρ_{cc} = 20%								
Ceiling		80%			50%			10%		
Walls		50%	30%	10%	50%	30%	10%	50%	30%	10%
Room Cavity Ratio	1	73	71	68	66	46	62	57	56	55
	2	65	60	56	58	55	52	51	49	47
	3	57	52	48	52	48	45	46	43	41
	4	51	45	41	47	42	39	41	39	35
	5	46	39	35	42	37	33	37	33	31
	6	41	35	31	37	33	29	33	30	27
	7	37	31	27	34	29	25	30	26	24
	8	33	27	23	30	25	22	27	23	21
	9	30	24	20	27	22	19	24	21	28
	10	27	21	18	25	20	17	22	18	16

Maximum recommended spacing-to-mounting height ratio above work plane is: 1.4

Figure 5-11.   Standard Photometric Test Report

6

interior lighting and the luminous environment

Artificial lighting involves an interaction of luminous energy, the room, and the visual task that permits an individual to perform a visual task in a given environment. So far, the source of luminous energy, lamp and luminaire, has been discussed. The room is as important in artificial lighting as the task and luminaire.

IMPACT OF THE ENVIRONMENT (ROOM)
ON LUMINAIRE SELECTION

Seven characteristics of a room have an impact on the lighting:

1. Surface reflectances
2. Size of room
3. Shape of room
4. Windows
5. Maintenance
6. Temperature
7. Furniture or equipment in the environment

Windows and their effect are discussed in Chap. 10; the effect of temperature on light sources is discussed in Chap. 4.

Surface Reflectances and Room Proportions

The light that falls on any seeing tasks in a room comes from two sources, the luminaire itself and the surfaces of the room. The amount of direct light from the luminaires reaching the task is dependent on maintenance and temperature. The reflected component, however, is dependent on surface reflectance, location of the luminaire, and size of the room. Each surface in a room (walls, ceiling, and floor) can have a different reflectance. The reflectance of each has a different effect on the contribution of reflected light to the illumination level produced in a room.

Depending on the type of lighting system used, the ceiling may be of major importance. With a totally direct lighting system, the amount of light reflected off other surfaces that can reach the ceiling is small. The reflectance of the ceiling for a direct system is less important in making a contribution to the work surface. However, ceiling reflectance is important in reducing luminance ratios between the luminaire and the ceiling. The higher the ceiling reflectance, whether a direct or indirect system is used, the lower the luminance ratios and the more comfortable the occupants will be.

The coefficient of utilization (CU) which is defined in Chap. 7, can be used as a measure of the efficiency of a luminaire in getting luminous energy to the work surface in a particular room. The CU value combines the room characteristics with the luminaire characteristics to indicate the efficiency of the interaction of the two components. Figure 6-1 shows the effect on the coefficient of utilization for a direct and direct–indirect luminaire (see Chap. 5 for definitions) for a 30 percent change in ceiling reflectance. With the direct–indirect luminaire, the ceiling reflectance becomes more important because of the luminous energy that goes directly to the ceiling; this results in a greater space between the 50 and 80 percent ceiling curves. With the direct luminaire, the difference between the 50

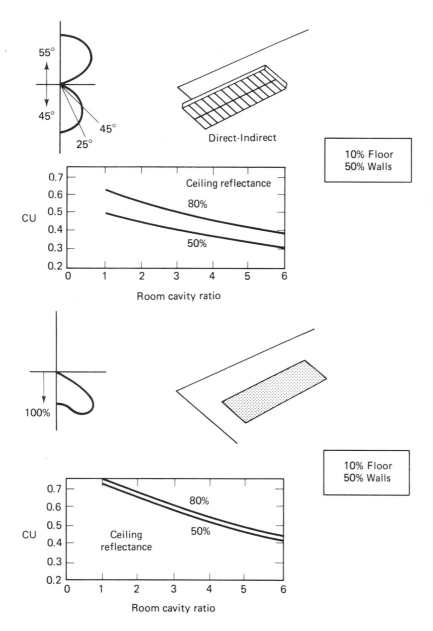

Figure 6-1. Effect of Ceiling Reflectance on the Coefficient of Utilization

and 80 percent ceiling curves is less, which indicates the lesser role the ceiling plays in utilizing luminous energy.

The amount of light falling on the walls in a good lighting system should be of sufficient magnitude to provide field luminances and luminance ratios that are comfortable. Depending on reflectance properties, the wall luminance and thus the interreflected component will change. Figure 6-2 shows the effect on CU with

140

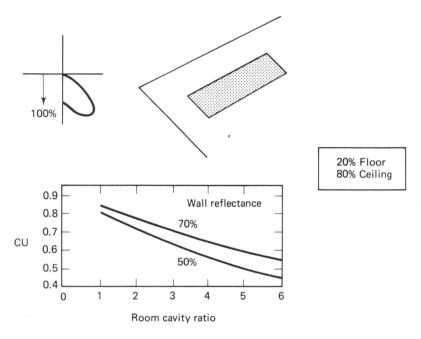

Figure 6-2. Effect of Wall Reflectance on the Coefficient of Utilization

a direct luminaire when the wall reflectance is increased by 20 percent. This shows that, ideally, all-white walls would give the highest illumination level. The sterile environment of all white can cause occupant distress, and color is necessary to provide psychological relief. Any color has a reflectance less than white, but the proper choice of colors will provide a pleasant, well-lighted environment.

Even if the architect or interior designer provides light-colored walls, the owner may reduce the efficiency of the system by installing dark furniture, pictures, blackboards, and the like, that absorb light. If energy conservation is to be the responsibility of the design team, they must have a voice in the selection of both the reflectance and furniture, and the placement of both.

The floor, depending on the particular tasks involved, can be important in providing reflected light. Table 7-5 shows that the floor correction factors for 10 or 30 percent floor reflectances have little effect on the CU value. For tasks located on top of the work plane, there is little effect. In this situation, the light has to be reflected off the floor, to the ceiling, and finally to the task. If the task is oriented on the bottom of the work plane, on the other hand, the floor reflectance becomes important. An example is in the design of hangars for repair work on airplanes. Many of the tasks are on the underside of wings or fuselage. With a low-reflectance floor, supplementary lighting is required. Bringing the floor reflectance up from 20 to 80 percent may eliminate the need for supplementary lighting.

The size and shape of the space affect the distribution of luminous flux in the space. To show the effect, the coefficient of utilization of the zonal cavity

141

method will be compared for changes in geometry. A review of the photometrics of luminaire in the data section of Chap. 7 will show that the CU decreases as the room cavity ratio (RCR) increases. For rectangular rooms with the same floor area and same room cavity height, a change in the proportion of the length and width does not affect the RCR. A decrease in wall area (lower mounting heights) to floor area increases light utilization; thus the RCR is lower. The larger the RCR, the lower the CU. This is logical, because the increase in distance from the luminaire to the work plane decreases the illumination by virtue of the inverse-square law. A change in shape, from a rectangle to a circle with the same floor area, shows an interesting decrease in the RCR. A circular room has a higher CU value than a rectangular room of the same floor area, and thus a higher illumination level. If the floor area is increased while the room cavity height remains the same, the RCR decreases. That is, large rooms can be lighted more efficiently.

Maintenance

A chief factor affecting the performance of a lighting system is the deterioration of the components. The accumulation of dirt, lamp depreciation, and discoloration of lenses decreases light output. These enter into the lumen formula of Chap. 7 as lamp lumen and luminance dirt depreciation. Other factors include luminaire surface depreciation, burnouts, and room surface dirt depreciation.

The design of a lighting system should be such that the required illumination is provided when the system is putting out the least amount of light. Since a decrease in efficiency does occur, this provides the occupants with at least the required illumination. The extent to which the system is overdesigned at the start is dependent on how well the maintenance program is carried out once the system is installed. If dirt is allowed to accumulate, lamps are not replaced until they burn out, and discolored lenses are not replaced, the lighting system is going to require a larger initial investment to compensate for light loss.

Lamp Lumen Depreciation (LLD)

The light output of lamps decreases as they get older. This is known as lamp lumen depreciation. The values for different lamps are given in the data section at the end of Chap. 7. The effect of lumen depreciation can be reduced by planned replacement, such as group relamping where all the lamps are replaced at some interval less the 100 percent of rated life. Generally, group replacement of lamps is more economical than the spot relamping of individual burnouts. An analysis of the situation will determine the most economical method based on costs, desired illumination, and replacement of burnouts (those that burn out before the scheduled relamping). By using the following equations, the cost per socket for lamps and labor for a year can be found.

Individual replacement:

$$\text{dollars/socket/year} = \frac{B}{R}(c + i)$$

Group replacement (early burnouts replaced):

$$\text{dollars/socket/year} = \frac{B}{A}\left(c + g + KL + Ki\right)$$

Group replacement (no replacement of early burnouts):

$$\text{dollars/socket/year} = \frac{B}{A}\left(c + g\right)$$

where B = burning hours per year
R = rated average lamp life, years
A = burning time between replacements, hours
c = net cost of lamps, dollars
i = cost per lamp for replacing lamps individually, dollars
g = cost per lamp for replacing lamps in a group, dollars
K = proportion of lamps failing before group replacement (from lamp mortality curve, Figure 6-3)
L = net cost of replacement lamps, dollars

The variables in the equations can take on different values depending on historical information or the estimate of the engineer. Burning hours, B, can be found from the typical operation of the facility. For example, an office area could be in use for 2340 h/yr (9 h/day × 5 days/wk × 52 wk/yr). The rated average lamp life, R, is found in manufacturers' catalogs. Currently, a fluorescent lamp has an R of 22,000 h. The net cost of lamps, c, depends on the quantity

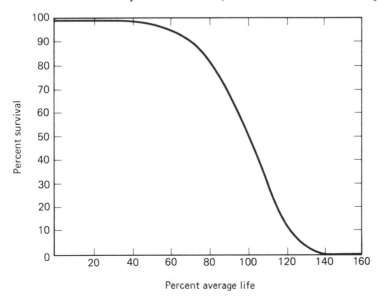

Figure 6-3. Lamp Mortality Curve

purchased. A fluorescent F40T12 lamp is about $0.93 per lamp. Records or time studies will determine the labor cost involved in individual replacement, *i,* and group replacement, *g,* of each lamp. With a fluorescent system, the replacement of individual lamps could cost $10 or more. If all four lamps were replaced in a four lamp luminaire, the cost per lamp could drop. A possible figure would be one fourth the individual replacement cost, since only one setup would be made to change all four. In a program where early burnouts are replaced before group relamping, either new lamps or lamps saved from a previous group relamping can be used to replace those that have burned out. The replacement lamp cost, *L,* can then be from the cost of a new lamp, *c,* to nothing. Typically, spot replacement is done with used lamps, and the cost, *L,* is zero.

The above factors depend, for the most part, on easily identifiable quantities. The burning time between replacement, *A,* and the proportion of lamps failing before group replacement, *K,* are dependent on the desired maintained illumination. For example, if early burnouts are replaced, the time between group replacement, *A,* can be longer than if they are not replaced. For the particular system under investigation, it has been decided that 10 percent can fail with no replacement before group replacement is done. This is about 70 percent of average life (Figure 6-3), or approximately 14,000 h between replacement for fluorescent lamps. If burnouts are replaced, the decision may be to allow as many as 25 percent to fail. This raises the time between group replacement to about 18,700 h.

Using the numbers in the previous discussion, it is determined that group replacement with no early burnout replacement is the most economical method.

Individual replacement:

$$\text{dollars/socket/year} = \frac{2340}{22,000}(0.93 + 10.00)$$

$$= \$1.63/\text{socket/year}$$

Group replacement (early burnouts replaced):

$$\text{dollars/socket/year} = \frac{2340}{18,700}[0.93 + 2.50 + (0.25 \times 0) + (0.25 \times 10.00)]$$

$$= \$0.742/\text{socket/year}$$

Group replacement (no replacement):

$$\text{dollars/socket/year} = \frac{2340}{14,000}(0.93 + 2.50)$$

$$= \$0.573/\text{socket/year}$$

This analysis shows which method is least costly. There is no formula to determine the time between replacements, *A,* and the proportion failed before replacement, *K*. Various values, chosen for their effect on the illumination, should be tried to find the alternative with the least cost.

Luminaire Dirt Depreciation (LDD)

Airborne dirt deposited on luminaire surfaces affects the light output and distribution. The environment, that is, the amount and kind of dirt, determines dirt depreciation. Also, the luminaire design, lamp, and finish can determine how much of the dirt will eventually deposit on the luminaire. Ventilated designs tend to collect dirt less rapidly than those with closed tops. On the other hand, completely closed units, if properly designed, exclude almost all dirt (see Figure 6-4).

Luminaire surfaces themselves can also lose their ability to reflect or transmit light. Anodized and enameled reflective surfaces deteriorate. Environments that contain highly caustic materials cause surfaces to deteriorate faster. The use of plastic lenses has presented the added problem of yellowing. Depending on the light source, luminaire temperature, and type of plastic, the period of time before the transmitting properties of the lens begin to deteriorate varies.

Luminaire cleaning at the time of lamp replacement brings the system to optimum efficiency. By relamping and cleaning at the same time, labor costs can be cut. Lenses should be replaced whenever they yellow, although nothing can be done about luminaire surfaces that have deteriorated except to replace the entire luminaire.

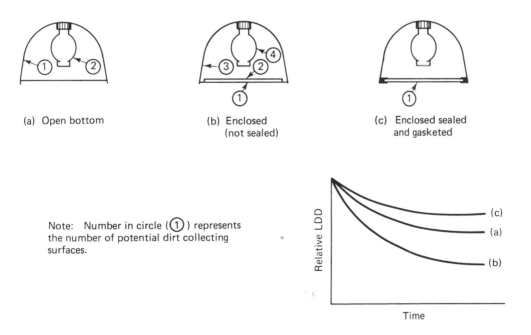

(a) Open bottom

(b) Enclosed (not sealed)

(c) Enclosed sealed and gasketed

Note: Number in circle (①) represents the number of potential dirt collecting surfaces.

Figure 6-4. Luminaire Dirt Depreciation

Room Surface Depreciation (RSD)

It would seem that, if the luminaires are cleaned and relamped occasionally, the major light loss factors will have been taken care of. However, since high room surface reflectances have been shown to be a requirement for good system efficiency, any decrease in room surface reflectances will also increase light loss. Depending on the environment, room surfaces should be washed periodically and painted when the reflectance has dropped significantly. Maintenance should be performed more frequently when a larger percentage of light is reflected by these surfaces. It should be noted that sometimes surfaces appreciate; that is, reflectance rises. Bleaching out of paint, curtains, or other materials, as well as light-colored dust, can cause the appreciation of surface reflectances.

Objects in the Environment

Most preliminary design assumes that the space is entirely empty. When furniture and equipment are placed in a room, the distribution of the light can change radically. Furniture that is light in color with a matte finish will prevent absorption of light and glare. Shadows that occur from equipment shielding light can be prevented by properly locating equipment with respect to the lighting system. As task-lighting equipment (i.e., equipment that is located near the task and oriented properly) is improved, the problem with shadowing will be minimized.

IMPACT OF THERMAL AND ACOUSTICAL CHARACTERISTICS OF LUMINAIRE ON THE ENVIRONMENT

Besides putting light into the space, luminaires introduce thermal and acoustical changes to the space. Most of the time, lighting designers ignore the impact of lighting equipment on the environment. To provide an integrated environment, all factors must be considered.

Thermal

Electrical equipment (including lighting) produces 3.41 Btu of heat per hour per watt of power. The transfer of this heat energy occurs by radiation, conduction, and convection. This heat can be used in the winter to supplement heat requirements, but must be removed when cooling is required. This load on the air-conditioning system is taken care of by increasing the capacity of the cooling equipment. It has been suggested that energy can be saved by turning off lights in existing installations, resulting in a savings in the energy used for lighting as well as the energy used for cooling. In most systems this may be true; however, in the case of systems using electric resistance heaters to maintain the design air temperature, reduction in air temperature causes the reheat coils to cycle on, increasing the air temperature to its design level. Therefore, each system must be analyzed carefully to determine the impact of lighting on the mechanical system.

The coordination of lighting and mechanical design is essential to the optimization of both.

Selection of efficient sources and luminaires, proper room surface reflectances, and the use of task lighting will reduce the lighting system wattage. Any heat produced must still be considered in the design of the mechanical system.

If the heat from the lighting system can be utilized rather than being dumped or wasted, the energy consumed by the heating system can be reduced to the degree to which the lighting provides heat to the space. The heat from the lighting system must be redistributed to areas where it is needed. The common means of using the heat produced by the lighting is the heat-transfer luminaire.

Heat-Transfer Luminaire

To control heat gain due to lighting energy, one can use heat-transfer luminaires. The two basic heat-transfer mechanisms used today are air and water.

Air is essential to proper ventilation of a space and is useful as a heat-transfer medium. The air circulation around and through the luminaire is essential for maximum heat transfer. However, excessive air-temperature differences and air motion can affect the light output and color of fluorescent lamps (see Chap. 4) due to pressure changes inside the lamp.

Circulating water is also an effective heat-transfer mechanism. The primary advantage is that it requires substantially less volume for the same heat removal capacity. The use of water to remove luminaire heat does not eliminate the need for air systems. Air is essential to satisfy humidity control, air motion, and ventilation. The volume of air will be less in the water system as compared to the all-air system.

The use of infrared reflecting filters in the lens will result in preventing infrared energy from entering the occupied space. This means that the heat removal system in the luminaires must be very efficient to remove this added heat.

The control of heat produced by lighting can lower heat gain in the surrounding space, reduce air changes (reduce fan horsepower), lower temperature differentials, reduce luminaire and ceiling temperature, and increase luminaire efficiency and life. The increase in initial cost due to heat-transfer luminaires may be offset by the reduced cost of lamps and the reduction in the mechanical system.

Temperature and Lamp Lumens

A secondary benefit of the heat-transfer luminaire is the potential for increased light output. Chapter 4 shows that a variation in ambient temperature can affect the light output of a fluorescent lamp. The buildup of heat can decrease light output. The heat-transfer luminaire, if properly designed, can reduce the ambient air temperature in the luminaire, resulting in more efficient operation and utilization of energy. Proper luminaire housing design and mounting details conduct some of the heat away, but not as much as the heat-transfer luminaire. As energy becomes more expensive and scarce, every Btu will have to be used efficiently, which means a more thorough investigation of the interaction of the system.

Acoustical

In the past few years, concern about the acoustical environment has increased as new technology has produced an increasingly noisy environment. The lighting system has also added to the noise pollution that surrounds us. Lighting equipment, especially gaseous discharge lamps, which require inductance ballasts, is a noise source. Incandescent lamps, if dimmed, may emit an audible ringing owing to filament vibration; the noise can be eliminated by adding a choke coil in the power supply line. Fluorescent ballasts have undergone numerous improvements, including a reduction in noise level. They are available for very quiet installations (A rated), to those where a high noise level is acceptable (F rated) (see Table 6-1). Fluorescent equipment manufactured for indoor applications should be equipped with an A sound rated ballast. Even with an A rated ballast, the fluorescent luminaire will emit some acoustical energy. A poorly constructed luminaire or a poor mechanical connection of the ballast to the luminaire will cause a speaker effect. If the luminaire is well constructed and noise still presents a problem, the ballast can be located outside the critical area. This can be expensive and can create a voltage drop, which should be investigated in advance.

TABLE 6-1

Fluorescent Ballast Sound Ratings

Sound ratings	Average noise rating (dB)
A	20–24
B	25–30
C	31–36
D	37–42
E	43–48
F	49 and up

Unlike fluorescent ballasts, the other gaseous discharge ballasts have been used almost exclusively outdoors where a high noise level could be tolerated. With the energy crisis, more indoor applications of HID equipment have arisen. There is no standard industry sound rating for HID ballasts. The noise characteristics of HID ballasts used indoors should be evaluated by the designer before they are specified. As with fluorescents, luminaire construction can have as much effect on the sound level as the ballast alone.

A possible solution to the noise problem is the use of a high-frequency solid-state ballast, which has been under development for some years. Such a ballast would add significantly to the advantages of using HID luminaires indoors. The advantages of high frequency solid state ballasts are its light weight, almost no noise, increased lamp efficacy, increased lamp life, simpler controls, and less heat.

The noise generated by a luminaire is not the only effect the lighting system can have on the acoustical environment. Most lighting equipment is installed in the ceiling where acoustical tile absorbs sound. The hard surfaces of the luminaire on the other hand reflect sound. In an open office where partial-height partitions are used for privacy, an improperly located luminaire can reflect speech from one cubical to the other, thus negating the purpose of the partition. If the walls extend only up to the suspended ceiling, the plenum will be completely open. An opening in the luminaires will allow sound to travel through one luminaire into the plenum and out another luminaire.

IMPACT OF THE DISTRIBUTION OF LUMINAIRES ON THE ENVIRONMENT

The distribution of luminaires can be described in terms of (1) the luminous output of a single luminaire, (2) the luminous output of the luminaires in the environment, and (3) the location of the luminaires in the environment. The luminous output of a single luminaire is characterized by the luminous intensity (candlepower) distribution. Chapter 5 shows the calculation of zonal lumens, which are derived from luminous intensity. The distribution of zonal lumens in the upper and lower hemispheres serves as the basis for the CIE–IES luminaire classification (see Chap. 5), which describes the general characteristics of the luminous output of a single luminaire.

The distribution of luminaires (output and location) in the environment can be classified as (1) general lighting, (2) localized lighting, and (3) localized-general lighting. Each of these classifications will be described in terms of the luminous output in the environment and the location of the equipment.

General Lighting (Ambient)

General lighting produces a uniform level of illumination throughout the environment. The uniform level of illumination is usually achieved by a symmetrical placement (layout) of luminaires in the space. Classically, this has been the approach used to solve most lighting problems. Since worker locations were not known, the lighting engineer put task levels of illumination throughout the entire space. This approach resulted in spaces being overlighted, but no one objected since power was abundant and energy was cheap. This method is not consistent with today's declining resources, which resulted in power cutbacks and rapidly rising energy cost.

The concept of general lighting is not dead. The general lighting approach has simply changed directions from high (task level) uniform light levels to lower uniform light levels. The term **ambient lighting** has been coined to describe the use of lower uniform (general) lighting levels throughout the space. Ambient lighting reduces luminance ratios in the visual field to maintain visual comfort. It also provides a general level of illumination for the safe circulation of people in the space. The currently accepted standard is to use ambient lighting levels that produce acceptable luminance ratios within the visual field for the particular application. Ambient lighting can be provided by two modes of application: (1) direct and (2) indirect.

Direct Mode

The term "direct" in this sense is used to describe the location of the lighting equipment. The direct mode implies that the luminaires are attached to the ceiling and most, if not all, of the luminous intensity is directed down toward the work surfaces. Luminaires classified as direct (CIE–IES classification) are most commonly used in this mode of application because of their greater efficiency (higher CU) and lower initial cost. However, CIE–IES classifications of semi-direct and direct–indirect can be used to satisfy the direct mode of application. A list of advantages and disadvantages is given in Table 6-2 to help the lighting engineer evaluate the direct mode of application. The list may not be complete, but is given to encourage in-depth analysis of the lighting system put into spaces.

TABLE 6-2

Analysis of Direct (CIE-IES Classification) Lighting

Advantages	*Disadvantages*
More energy efficient	Harsh shadows
Low initial cost	Greater potential for direct glare
Lower floor to ceiling height	Greater potential for veiling reflections and reflected glare
Easier to control and direct	Excessive luminance ratios on the ceiling
	Excessive luminance and glare causes a change in focal point to the ceiling

A lighting system placed in or on the ceiling is intended to produce illumination for good visibility. Owing to their potentially excessive brightness and glare, direct-mode systems can create a focal point on the ceiling plane rather than at other more important areas within the environment. That is, the focal point(s) created by the lighting system may become a visually dominant feature in the space. In most applications this is not the function of the lighting system. The lighting system should be visually subordinate to other aspects of the environment.

The use of the direct mode of application, along with localized lighting, is a viable means of solving certain types of lighting problems. Through the proper selection of a direct-mode system that controls glare, the overall visibility, visual comfort, and efficiency of the lighting system can be improved.

Indirect Mode

Indirect mode implies that most, if not all, of the luminous intensity is directed off the ceiling and arrives on the work plane indirectly. The luminaires can be floor mounted or suspended from the ceiling. The luminaires can be CIE-IES classification semi-indirect or indirect. Since the ceiling becomes the primary source of light relative to the work plane, the ceiling surface should be uniform (unobstructed), highly reflective, and diffuse in character. These characteristics will improve the visual appearance of the ceiling plane and, if combined with a properly designed indirect lighting system, will result in uniformity of the ceiling. A list of a few of the advantages and disadvantages of an indirect lighting system is provided in Table 6-3 for analysis of the indirect mode.

TABLE 6-3

Analysis of Indirect (CIE–IES classification) Lighting

Advantages	*Disadvantages*
Potential for maximum visual comfort	Less energy efficient
Higher ESI	Requires a uniform, unobstructed, diffuse reflecting ceiling surface
Greater direct glare control	Increase in building cost, due to increased floor-to-ceiling height for good uniformity
Blends in with the environment, reducing the potential focus point change	National Electric Code requirements and electrical distribution system may increase cost
Produces uniform, diffuse, shadow-free light	
Qualifies as tangible personal property (a bonus for depreciation and investment tax credit) if light is part of the furniture	

A misleading claim and common misconception relative to the indirect lighting system is that better energy utilization is possible with indirect lighting systems. Since a coefficient of utilization represents the interaction of the luminaire distribution with the room characteristics, a standard coefficient of utilization can be used as an indicator of the overall system efficiency within a given environment. Table 6-4 shows us that the indirect system can never be more energy efficient than a direct system. Table 6-4 also shows the change in coefficient of utilization for a typical fluorescent luminaire as we change the luminaire from an indirect application to a direct application.

Once the indirect mode has been selected as a means of providing ambient lighting, the question then becomes one of choosing the light sources to be used in the indirect application. A number of systems utilize fluorescscent lighting; others use HID lamps, such as phosphor-coated mercury, metal halide, or high-pressure sodium lamps.

The use of HID lamps in an indirect application will result in better energy utilization (Table 6-4) than a fluorescent system. When the control characteristics

TABLE 6-4

Typical Luminaire Coefficient of Utilization: Direct and Indirect Mode

FLUORESCENT:
Coefficient of Utilization
Prism-K12

RCR	Up 0.70				Down 0.70			
	0.70	0.50	0.30	0.10	0.70	0.50	0.30	0.10
1.0	0.48	0.46	0.44	0.42	0.70	0.68	0.66	0.64
2.0	0.43	0.40	0.37	0.34	0.65	0.61	0.58	0.55
3.0	0.39	0.35	0.31	0.29	0.60	0.55	0.51	0.48
4.0	0.36	0.31	0.27	0.24	0.56	0.50	0.46	0.42
5.0	0.33	0.27	0.23	0.20	0.52	0.45	0.40	0.37
6.0	0.30	0.24	0.20	0.18	0.48	0.41	0.36	0.33
7.0	0.28	0.22	0.18	0.15	0.44	0.37	0.32	0.29
8.0	0.26	0.20	0.16	0.13	0.41	0.33	0.29	0.25
9.0	0.24	0.18	0.14	0.12	0.38	0.30	0.25	0.22
10.0	0.22	0.16	0.13	0.10	0.35	0.27	0.23	0.19

HID:
Coefficient of Utilization
Bollard

RCR	Up 0.70				Down 0.70			
	0.70	0.50	0.30	0.10	0.70	0.50	0.30	0.10
1.0	0.59	0.56	0.54	0.52	0.81	0.77	0.72	0.69
2.0	0.54	0.49	0.46	0.43	0.70	0.62	0.55	0.49
3.0	0.49	0.43	0.39	0.35	0.61	0.51	0.42	0.36
4.0	0.44	0.38	0.33	0.30	0.54	0.43	0.34	0.27
5.0	0.41	0.34	0.29	0.25	0.48	0.36	0.27	0.20
6.0	0.37	0.30	0.25	0.22	0.43	0.30	0.22	0.16
7.0	0.34	0.27	0.22	0.19	0.39	0.26	0.18	0.12
8.0	0.32	0.24	0.20	0.16	0.36	0.23	0.15	0.10
9.0	0.29	0.22	0.17	0.14	0.33	0.20	0.13	0.08
10.0	0.27	0.20	0.16	0.13	0.30	0.18	0.11	0.06

of a fluorescent luminaire versus a HID luminaire are studied, it becomes apparent that a properly designed HID unit will always be more efficient and easier to control than a fluorescent unit. Some of the lower-mounted indirect fluorescent lighting systems (at or below eye level) utilize a **low-brightness** louvered system to control the direct glare. These systems are less efficient than other forms of control systems, such as lenses (see Table 6-4). A comparison of the indirect coefficients of utilization for a HID system (see Table 6-4) to the direct application of the low-brightness louvered system (see Table 6-5) shows the indirect application of the HID system has a higher coefficient of utilization than the direct application of the low-brightness fluorescent system.

In addition to the selection of the light source, another critical factor in the indirect system is the distance between the luminaire and the ceiling plane. To provide maximum spacing and good uniformity, the indirect units must produce a wide distribution of luminous intensity or a batwing. Improperly or inadequately designed luminaires will create a spotty pattern on the ceiling surface. This will result in a change in focal point back to the ceiling surface. If the ceiling is the focal point, the ceiling will again become a visually dominant feature in the space, which makes the system less attractive than the direct mode.

The goal of the indirect mode should be to minimize or eliminate the spotty appearance of the ceiling surface. This requires two things in the indirect system: (1) a properly designed optical system to provide a uniform illumination level across the surface of the ceiling, and (2) proper selection of a ceiling material. Ideally, the ceiling surface should be perfectly diffuse or approach a perfect diffuser. This requires a matte, high-reflectance surface with a very slight texture to help diffuse the light on the surface. Exposed T-bar ceiling systems can create problems of reflected images off the glossy, painted surface of the T-bar.

The luminaire should be at a sufficient distance off the floor to prevent direct viewing of the lamp and housing. If the luminaire housing is below eye level, the more inefficient low-brightness systems must be used to control the glare.

A well-designed indirect lighting system will produce uniform luminance across the ceiling surface and uniform, diffused, shadow-free light throughout the space at the work level. Uniform diffused, shadow-free light is quite good from a visual standpoint. Aesthetically and psychologically, uniform, diffuse, shadow-free illumination can be quite boring and uninteresting. The lighting engineer should take this into consideration and provide visual relief from the boredom of the uniform diffused illumination.

Localized Lighting

Localized lighting is achieved by placing the lighting in the area immediately over the task. Spill light from the luminaire is expected to produce some general illumination in the space. Localized lighting alone is normally not adequate to minimize luminance ratios in the environment. Visual discomfort may result; therefore, this type of lighting system should be avoided in areas where critical visual tasks are performed.

TABLE 6-5

Typical Low-Brightness Fluorescent System: Direct and Indirect Modes

Coefficient of Utilization
Plastic louver

RCR	Up 0.70				Down 0.70			
	0.70	0.50	0.30	0.10	0.70	0.50	0.30	0.10
1.0	0.37	0.35	0.34	0.33	0.54	0.52	0.51	0.49
2.0	0.34	0.31	0.29	0.27	0.50	0.47	0.44	0.42
3.0	0.31	0.27	0.24	0.22	0.47	0.42	0.39	0.37
4.0	0.28	0.24	0.21	0.19	0.43	0.38	0.35	0.32
5.0	0.25	0.21	0.18	0.16	0.40	0.35	0.31	0.28
6.0	0.23	0.19	0.16	0.14	0.37	0.32	0.28	0.25
7.0	0.21	0.17	0.14	0.12	0.34	0.29	0.25	0.22
8.0	0.20	0.15	0.12	0.10	0.32	0.26	0.22	0.20
9.0	0.18	0.14	0.11	0.09	0.30	0.24	0.20	0.17
10.0	0.17	0.13	0.10	0.08	0.28	0.22	0.18	0.16

Coefficient of Utilization
Low-brightness wedge

RCR	Up 0.70				Down 0.70			
	0.70	0.50	0.30	0.10	0.70	0.50	0.30	0.10
1.0	0.33	0.31	0.30	0.29	0.49	0.48	0.47	0.46
2.0	0.30	0.27	0.25	0.24	0.46	0.44	0.42	0.41
3.0	0.27	0.24	0.21	0.20	0.44	0.41	0.39	0.37
4.0	0.25	0.21	0.18	0.16	0.41	0.38	0.35	0.33
5.0	0.22	0.19	0.16	0.14	0.38	0.35	0.32	0.30
6.0	0.21	0.17	0.14	0.12	0.36	0.32	0.29	0.27
7.0	0.19	0.15	0.13	0.10	0.34	0.30	0.27	0.25
8.0	0.17	0.13	0.11	0.09	0.31	0.27	0.24	0.22
9.0	0.16	0.12	0.10	0.08	0.29	0.24	0.21	0.19
10.0	0.15	0.11	0.09	0.07	0.27	0.22	0.19	0.17

Localized-General Lighting (Task/Ambient)

Localized-general lighting can be achieved by providing a uniform general lighting level throughout the space and localized lighting where higher task illumination levels are required. This type of localized lighting is called **task lighting**. The general lighting requirements can be satisfied by either direct- or indirect-mode applications.

A second approach to localized-general lighting would consist of a functional arrangement of luminaires relative to the task location. The nonsymmetrical arrangement of luminaires must provide adequate general illumination. This will result in a nonuniform general illumination in the space and a nonsymmetrical luminance pattern (luminaires or ceiling) on the ceiling plane. Nonsymmetrical patterns from direct- or indirect-mode applications can draw attention to the ceiling, possibly resulting in a visually dominant ceiling.

Task lighting can be provided by overhead luminaires mounted in the ceiling or luminaires mounted near the task area in a furniture module.

Furniture Task Lighting

Furniture task lighting has the advantage of a potential decrease in power consumption due to the decreased distance from the luminaire to the task. Because of the close proximity of the lighting equipment and the confined area of the furniture module, the possible luminaire locations are limited. The easiest location is directly in front of the task under a shelf or mounted on a backboard. This location will usually result in decreased visibility and visual discomfort, which may lead to an increase in the time required to perform visual work and thus an increase in building usage and energy consumption. Comfort control of task lighting is dependent on luminance ratios (1) within the furniture module, (2) between the luminaire and its immediate surround, and (3) between the furniture module and the far surround. Even though illumination levels (classical foot-candles) may be more then adequate, improperly located lighting systems relative to the task and the viewing position can result in a reduction in contrast, and hence visibility. This reduction in contrast may cause the user to sacrifice either speed or accuracy in performing work. As visibility decreases, the user will subconsciously either decrease his speed or lower his acceptance standards for accuracy. One hopes that the user, under those conditions, will make the proper decision as to which of these two aspects to sacrifice. The lighting engineer has available to him a technique by which he can evaluate the quality of a lighting installation in terms of visibility. How well do the people really see under the lighting system? The ESI system (see Chap. 7) is concerned with the reduction of task contrast and its effect on the performance of visual work.

Overhead Task Lighting

If task locations within the space can be defined, overhead task lighting has one major advantage: the entire ceiling is available for luminaire placement. This makes the control of visibility much easier as long as furniture locations do not change relative to luminaire positions. The relationship of luminaire location to task location is absolute and essential to maximum visibility. Because the luminaires are removed from the line of sight, the visual comfort potential is also quite good.

IMPACT OF THE LUMINOUS ENVIRONMENT ON THE USER

The goal of lighting design is to provide optimum visibility so that the user can perform work. Lighting is not placed in space simply to gratify the ego of a lighting engineer. Lighting systems that are put into critical visual areas for purely aesthetic reasons or for purely economic reasons, such as lower initial cost, may reduce visibility and create discomfort. This may result in emotional and psychological deprivation for the user, and an increase in the time spent to perform critical visual work.

references

1. Kaufman, John, editor, **IES Lighting Handbook,** Illuminating Engineering Society, New York, 1972.
2. Finn, J. F., "Efficient Application of Lighting Energy—A Luminaire Air Heat-Transfer Evaluation," **Lighting Design and Application,** Jan. 1976.

7

artificial illumination design techniques

Many techniques are used in artificial illumination design. Basically, all are variations of techniques that fall into two classifications, **quantity** and **quality.**

QUANTITY DETERMINATION METHODS

For many years only the quantity of illumination was considered. Although the quality of the illumination is considered today to be more important than the quantity, there are instances when the determination of the quantity, will suffice. The basic quantity technique in calculating illumination is the point-by-point method. Another useful method of determining the quantity of illumination is the lumen method.

Point-by-Point Method

The point-by-point method of determining the direct component of illumination is an application of the fundamental law of illumination, the inverse-square law:

$$E_n = \frac{I_\theta}{D^2} \qquad (1)$$

The law states that the illumination, E_n, at a point on a surface normal (perpendicular) to the light ray is equal to the luminous intensity, I_θ, of the source arriving at the point divided by the distance squared D^2, between the source and the point of calculation. These terms are shown in Figure 3-4.

Although the inverse-square law is simple, a number of assumptions limit its use. The most critical assumption is that the source must qualify as a point source. If the maximum source dimension is one fifth the distance to the point of calculation, the error will be less than 1 percent, and the source can be considered a point source. Because a point source is required, it may be necessary to divide a large area source into small segments. This is valid only if the luminance of the source is constant from point to point across the surface of the source at any viewing angle. Finally, asymmetrical distributions may cause error at particular points because the interpolation of luminous intensity between planes as given in published tables cannot be made reliably.

Horizontal Surface Illumination

The inverse-square law equation gives the illumination on a plane normal (perpendicular) to the direction of the incident light. When the plane of interest is at an angle to the incident light (Figure 7-1), the cosine law of illumination must be applied. The equation then becomes

$$E_h = E_n \cos \theta \qquad (2)$$

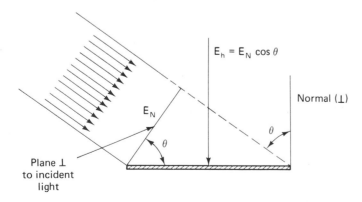

Figure 7-1. Cosine Law of Illumination

where E_h is the illumination on a horizontal plane due to luminous intensity incident at an angle θ to the normal. Replacing E_n in Eq. (2) with Eq. (1) gives

$$E_h = \frac{I_\theta}{D^2} \cos \theta \qquad (3)$$

If a number of calculations are to be made of illumination on a horizontal surface from an array of sources at the same height, H, the geometric relationship shown in Figure 7-2(a) is achieved.

$$D = \frac{H}{\cos \theta} \qquad (4)$$

can be substituted into Eq. (3) to give the cosine cubed law of illumination:

$$E_h = \frac{I_\theta \cos^3 \theta}{H^2} \qquad (5)$$

In this form, only the direction of the intensity I_θ need be known, eliminating the calculation of the distance to the point of interest.

Vertical Surface Illumination on a Vertical Line Through the Aiming Point. If the illumination on a vertical plane is desired, the sine of the angle is used as shown in Figure 7-2(b).

$$E_v = \frac{I_\theta}{D^2} \sin \theta = \frac{I_\theta}{D^2} \cos \phi$$

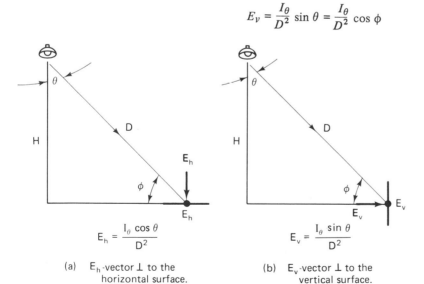

(a) E_h-vector $\perp$ to the horizontal surface.

$$E_h = \frac{I_\theta \cos \theta}{D^2}$$

(b) E_v-vector $\perp$ to the vertical surface.

$$E_v = \frac{I_\theta \sin \theta}{D^2}$$

Figure 7-2. Illumination on a Vertical and a Horizontal Plane

Vertical Surface Illumination on a Horizontal Line Through the Aiming Point. Equation (5) only applies for the calculation of illumination on a vertical surface along a vertical line through the aiming point. The vertical line is formed by the intersection of the vertical wall plane and a vertical plane through the luminaire that is perpendicular to the vertical wall plane. To find the illumination on the vertical wall plane at any other point requires the double trigonometric correction (see Figure 7-3).

$$E_v = \frac{I_\theta}{D^2} \sin \theta_s \cos \alpha \qquad (6)$$

The inverse-square law and the cosine law of illumination are basic to almost every method of illumination design. Isofootcandle diagrams and tables of illumination values are calculated by this design method. Only the direct component of illumination can be determined using the inverse-square law. Multiple interreflections from walls, ceiling, and floor may occur, making this method valid only for exterior calculations or in interior spaces where the interreflected component can be assumed to be negligible. A method of calculating the interreflected component is described in a report by the Committee on Lighting Design Practice of the IES[1].

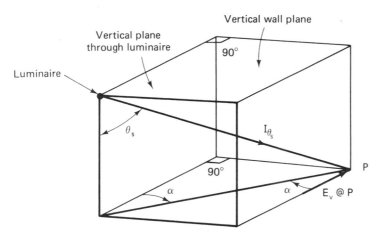

Figure 7-3. Vertical Illumination Off Axis

Lumen Method

Specifications often require the illuminating engineer to know or design for average uniform illumination. To do this with the inverse-square law for a large number of points would be both tedious and expensive. In addition, a second set of calculations would have to be made to determine the interreflected components. The lumen method was developed to simplify calculations and to determine average uniform illumination on a plane. The basic equation is

$$E_{ave} = \frac{(\text{lamps/LUM}) (\text{lumens/lamp}) (\text{no. of LUM}) (\text{CU}) (\text{LLD}) (\text{LDD}) (\text{RSD})}{\text{area WP}}$$

where WP = work plane
    LUM = luminaire
    CU = coefficient of utilization
    LLD = lamp lumen depreciation
    LDD = luminaire dirt depreciation
    RSD = room surface depreciation

The coefficient of utilization (CU) is calculated by the zonal cavity method and is a measure of how a specific luminaire distributes light into a given room. The CU takes into account luminaire efficiency, candlepower distribution of the luminaire, room size and shape, mounting height, and surface reflectances.

Coefficients of utilization for a particular luminaire are found in tables included as part of photometric tests for that luminaire. A typical example is shown in Table 7-3. CU values are also shown in Table 7-2 in the data section. The table assumes an effective floor cavity reflectance of 20 percent and a spacing-to-mounting height ratio of 0.40.

Figure 7-4 shows the parameters used to determine the CU in a room. The quantity known as the **cavity ratio** (CR) is defined in general terms as

$$CR = \frac{2.5 \ (\text{area of the walls})}{\text{area of the work plane}}$$

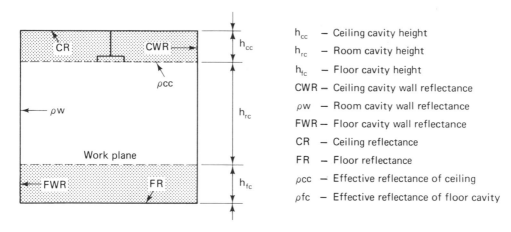

h_{cc} — Ceiling cavity height
h_{rc} — Room cavity height
h_{fc} — Floor cavity height
CWR — Ceiling cavity wall reflectance
ρw — Room cavity wall reflectance
FWR — Floor cavity wall reflectance
CR — Ceiling reflectance
FR — Floor reflectance
ρcc — Effective reflectance of ceiling
ρfc — Effective reflectance of floor cavity

Figure 7-4. Terminology Used in the Zonal Cavity Method

Each of the cavities formed by the luminaire plane and work plane has an associated cavity ratio:

$$\text{ceiling cavity ratio (CCR)} = \frac{2.5 h_{cc} \ (\text{perimeter of the walls})}{\text{area of the work plane}}$$

$$\text{room cavity ratio (RCR)} = \frac{2.5 h_{rc} \ (\text{perimeter of the walls})}{\text{area of the work plane}}$$

$$\text{floor cavity ratio (FCR)} = \frac{2.5h_{fc} \text{ (perimeter of the walls)}}{\text{area of the work plane}}$$

The height of the ceiling cavity is the suspension length of the luminaires; the floor cavity height is the distance from the floor to the work plane on which the average illumination is being calculated.

Referring to the flow chart (Figure 7-5), we need to know the effective ceiling cavity reflectance ρ_{cc}, the wall reflectance ρ_w, and an RCR to find a CU value. If the suspension length is zero, the effective ceiling cavity reflectance ρ_{cc} is equal to the actual ceiling reflectance. For suspension lengths other than zero, the ceiling cavity ratio (CCR) is calculated. Using Table 7-4 in the data section with the CWR, CR, and RCR, we can find an effective reflectance ρ_{cc}. The same procedure is used for an effective floor cavity reflectance ρ_{fc} when the work plane is any height off the floor.

With the effective ceiling cavity reflectance ρ_{cc}, the wall reflectance ρ_w, and the room cavity ratio (RCR) known, the CU is picked from Table 7-2. This CU value is based on a ρ_{fc} equal to 0.20. If the calculations have shown it to be something other than 20 percent, a correction factor must be applied to the CU. Correction factors can be found in Table 7-5 of the data section.

Since it would be impossible to generate tables to cover all reflectances, interpolation of the tables is required. Linear interpolation is sufficient for this type of calculation and is least time consuming. Refer to the example problems at the end of this chapter for the techniques involved. Additional examples can be found in reference 2.

Two other quantities needed in the lumen formula are the lamp lumen depreciation (LLD) and the luminaire dirt depreciation (LDD) factors. These factors take into account the decrease in light output of lamps over their life and the accumulation of dirt on luminaire surfaces. The LLD factors for different lamps are found in Table 7-7. To determine the LDD factor, each luminaire is categorized according to its maintenance characteristics. These categories are listed on the photometric data sheet (Table 7-2 and 7-3). Knowing the category, the appropriate graph (Table 7-6) is used to find the LDD factor. Assumptions must be made as to the cleanliness of the environment and the interval between luminaire cleaning.

The **zonal cavity** method is used to calculate a CU value that is used in the **lumen** method to calculate the average illumination on the work plane. To ensure that the illumination is uniform (a necessary condition in the lumen method), the recommended spacing to mounting height ratio (S/MH) should not be exceeded. In addition, to prevent drop off of illumination near the walls, the recommended spacing procedures shown in Table 7-8 should be followed.

A number of assumptions have been made in the zonal cavity method. The accuracy of the results is therefore limited. Also, it must be remembered that an average uniform illumination is calculated. Use of the lumen method should be limited to noncritical, uniform lighting and for making a preliminary determina-

involved is another factor in glare. Important elements that cause glare are size, luminance, and position of the light source and surround luminance. A small bright source directly on the line of sight against dark surroundings is an example of exaggerated glare.

Glare is complex and can be categorized into two major types, direct and indirect. **Direct glare** is due to excessively bright sources of light (luminaires and/or windows) in the field of view that shine directly into the eyes, resulting in discomfort and/or a loss in visibility; **indirect glare** is due to light sources that are reflected from tasks into the eyes, resulting in a loss in visibility. Direct glare is associated primarily with **heads-up** tasks (Figure 7-6); that is, the glare is produced by excessive luminances in the visual field that affect the visual system as the individual looks around the environment. There may also be a loss in visibility for vertical tasks in the heads-up situation. Indirect glare is associated with **heads-down** tasks. As the individual looks down at the work surface, indirect glare acts to create discomfort and/or a loss in visibility due to the action of the luminous environment on the task.

Relative to the luminaire, the direct-glare zone is associated with luminous intensity produced in the zone from 45 to 90°; indirect glare is associated with luminous intensity produced in the zone from 0 to 45° (Figure 7-6).

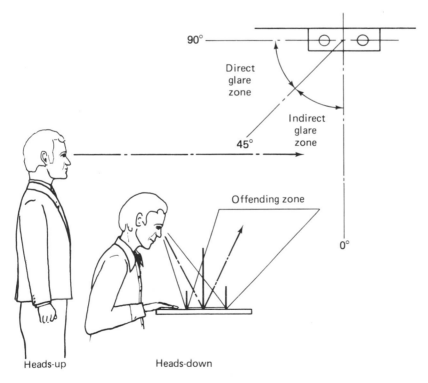

Figure 7-6. Direct and Indirect Glare

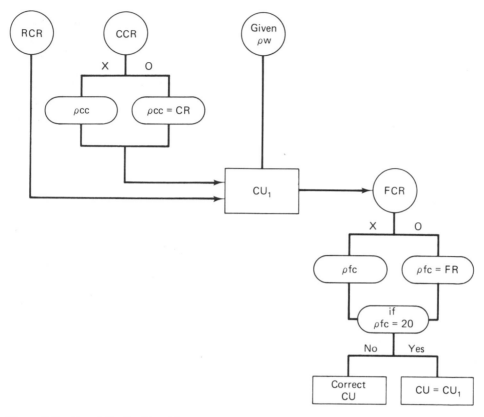

Figure 7-5. Zonal Cavity Flow Diagram

tion of the approximate number of luminaires needed. Where visibility of critical tasks is important, more sophisticated techniques should be used. The reader is referred to reference 1 for other techniques for determining the level of illumination.

QUALITY DETERMINATION METHODS

Glare

For a luminous environment to be visually comfortable, the occupants must not see any glare. **Glare** is defined as the result of excessive luminances in the field of view that are greater than the luminance to which the visual system is adapted. Physical discomfort (discomfort glare) or loss in visual performance and visibility (disability glare) can result from glare in the visual field.

Many factors are involved in producing glare. One is the length of time that the high luminance is present. Another factor is the luminance ratios between the glare source and the surround in the major portion of the field of view. The task

Direct glare (heads-up tasks) is evaluated in terms of visual comfort probability (VCP); indirect glare (heads-down tasks) is evaluated in terms of equivalent sphere illumination (ESI). Direct glare and indirect glare present two opposing problems; in general, to minimize direct glare, luminous intensity should be kept out of the 45 to 90° zone, which means that it is placed in the indirect-glare zone. To minimize indirect glare, the luminous intensity should be kept out of the 0 to 45° zone. That means it is placed in the direct-glare zone.

The designer must evaluate each room on the basis of the tasks performed to determine which type of glare is the greater problem. In general, it is difficult to optimize the solution to both direct and indirect glare, resulting in trade-offs.

Indirect glare involves two forms of glare: (1) reflected glare and (2) veiling reflections. **Reflected glare** is caused by specular or glossy surfaces reflecting images of bright sources into the eyes. Not as apparent as reflected glare, **veiling reflections** occur on surfaces and reduce the contrast of the task. The major problem with veiling reflections is that they are not visible; yet they reduce contrast and thus visibility.

Direct Glare (VCP)

Direct glare is largely dependent on the characteristics of the room and the light sources in the field of view. The Direct Glare Committee of RQQ Committee[3] of IES states that "direct glare may not be a problem in lighting installations if all three of the following conditions are satisfied":

1. The VCP is 70 or more.
2. The ratio of maximum-to-average luminaire luminance does not exceed 5 to 1 (preferably 3 to 1) at 45, 55, 65, 75, and 85° from nadir crosswise and lengthwise.
3. Maximum luminaire luminances crosswise and lengthwise do not exceed the following values:

Angle	Max. lum. (fL)
45	2250
55	1605
65	1125
75	750
85	495

An initial judgment as to the direct-glare potential of a luminaire can be made by examining the maximum and average luminances at the glare angles (45, 55, 65, 75, and 85°). A number of methods of evaluating direct glare have been devised. **Visual comfort probability** (VCP) is the method currently accepted by the IES for evaluating direct glare in a room. The VCP value represents the percentage of people that probably will not complain about the glare produced in the space.

The VCP method involves a luminaire-by-luminaire determination of the discomfort index for each luminaire within the field of view. The individual indexes are then combined in an appropriate manner to obtain an overall rating called the **discomfort glare rating** (*DGR*). To obtain the numerical VCP, the overall DGR rating is entered on a graph. Refer to reference 4 for a more detailed description of the VCP method.

Comfort varies widely with the individual. Therefore, the VCP rating system should act only as a guide to quality design. The VCP method assesses visual comfort that is related to discomfort (discomfort glare), which is commonly associated with direct glare.

The computation of VCP for a room is dependent on the following factors:

1. Room size and shape.
2. Surface reflectances.
3. Illumination level.
4. Luminaire type, size, and distribution.
5. Number, location, and orientation of luminaires.
6. Luminance of the entire visual field.
7. Observer location and line of sight.
8. Differences in individual glare sensitivity.
9. Equipment and furniture.

There are two ways to obtain a VCP for a room: by calculation or from tables. Naturally, the VCP at any point can be calculated, either by hand, using the method described by Guth[4], or by a computer program[5]. The second way is to obtain a VCP table from the manufacturer, who has already made calculations for a standard set of conditions and his luminaires.

Figure 7-7 shows a typical rectangular room. To facilitate comparison of the VCP calculated by different manufacturers, the following standard conditions have been established:

1. Initial illumination level of 100 fc (footcandles). (Number of luminaires determined by zonal cavity method.)
2. Surface reflectances: 80 percent for ceiling, 50 percent for walls, and 20 percent for floor.
3. Mounting heights of 8.5, 10, 13, and 16 ft.
4. A standard range of room dimensions.
5. Standard layout of luminaires uniformly distributed in the room. To meet this condition, the room is divided into 5- by 5-ft modules, and the total number of luminaires from step 1 is divided among the total number of modules (Figure 7-7).

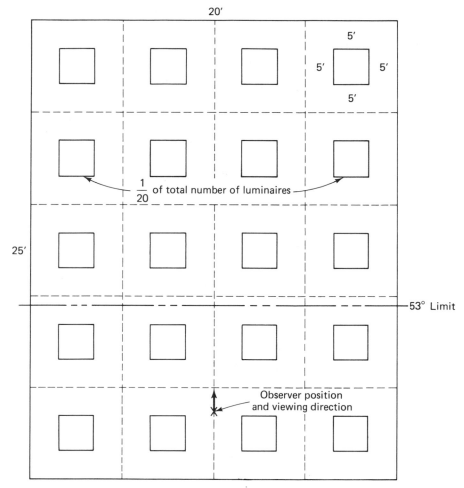

Figure 7-7. Typical Room for Calculating Visual Comfort Probability Tables.

6. An observation point that is assumed to be the worst location is selected to be 4 ft in front of the center of the rear wall and 4 ft above the floor.

7. A horizontal line of sight directly forward.

8. A limit of the visual field to 53° above and directly forward from the observer.

Standard tables can be used to make a preliminary evaluation of the effects of direct glare on visual comfort. However, the tables are too general to allow an accurate assessment of glare, especially if nonuniform lighting is utilized. The tables require uniform lighting layouts because of standard condition 5. The tables encourage non-energy-conserving uniform lighting layouts. Also, the standard observer location may not be the "worst" location or the actual task loca-tion[6].

Table 7-9 is a standard VCP Table. There are four different orientations possible in any given room. The width (W) column in the table is always the room dimension against which the observer's back is placed. Figure 7-8 shows the change in VCP for the luminaire in Table 7-9 as the luminaire orientation and observer location changes.

The preferred method for obtaining information on visual comfort is to analyze the actual conditions that exist in the luminous environment and obtain VCP values at specific locations within the space. Computerized techniques are available for assessing VCP values in a luminous environment for a specific layout for the lighting system.

Reflected Glare

Reflected glare results from specular reflections of high luminance areas off polished or glossy surfaces. Reflected glare may cause discomfort and, if reflec-tions are of high enough luminance and sufficient size, may also produce disability glare or a loss in visibility. Reflected glare is due to light that is reflected toward the eyes from glossy or semiglossy surfaces. The zone from 0° (relative to source) to 45° (Figure 7-6) is the most critical area for reflected glare.

The bright patches of light on the task are images of the light sources. The less specular or reflective the surface is, the less distinct the image will appear. The distinct image of a bright source is more annoying than the image of a diffuse source. Reflected glare is very distracting, annoying, a source of discomfort, and causes a reduction in visibility. The effects of reflected glare are more serious if the task surround is dark, owing to the luminance ratios created. Specular images in a highly polished dark walnut desk are much more harsh than they would be if reflected from a light-colored desk.

Veiling Reflections (ESI)

Veiling reflections veil the task with reflected light, resulting in a reduction of contrast. Veiling reflections are most commonly associated with light reflected off matte or diffuse surfaces. The reduction in contrast results in a loss in visibility and visual performance.

The contrast of a task is the difference in luminance between an object (print) and its background (paper).

$$C = \left| \frac{L_o - L_b}{L_b} \right| \quad (Blackwell\ Formula)$$

where C = contrast
L_o = luminance of the object
L_b = luminance of the background (see Figure 1-19)

The luminous environment interacts with the microstructure of the task, causing a change in the task contrast that affects the visibility of the task. This ef-

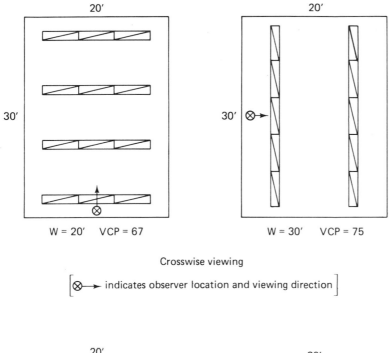

Crosswise viewing

[⊗→ indicates observer location and viewing direction]

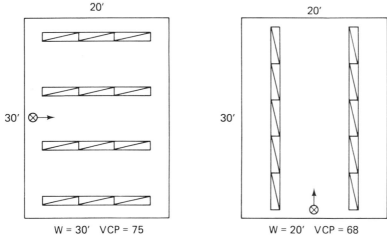

Lengthwise viewing

Figure 7-8. Application of Visual Comfort Probability Table in Data Section.

fect of veiling reflections on task contrast is evaluated in terms of equivalent sphere illumination (ESI). An understanding of the following example is sufficient for lighting engineers who must apply the ESI method.

Example　　　At a given viewing angle in a particular orientation, for a given task, in this luminous environment, at this point a value of 50 ESI fc is calculated. At the same point, an ordinary footcandle meter is placed and it reads 100 fc.

50 ESI footcandles—calculated
100 footcandles—measured or "classical"

What does this mean? If one were to take the task and place it inside a **photometric sphere,** it would require only 50 fc of illumination inside the sphere to produce **the same visibility** that the 100 fc produces in the real environment.

What is a photometric sphere? It is a large hollow sphere painted inside with a special high-reflectance, diffuse white paint. Light is introduced through an aperture in the side of the sphere. The light strikes a dispersion cone, which causes the light to reach the task after multiple reflections within the sphere. A second aperture is placed in the sphere at a given viewing angle to allow for the measurement of L_b and L_o produced in the sphere (see Figure 7-9).

A photometric sphere is used because

1. It provides a reference or standard lighting condition.
2. It is an easily reproducible standard.
3. It produces uniform diffuse light that is similar to the lighting conditions used in the basic research.

It is important to remember that a photometric sphere was **not** selected because it was **better** than any other lighting system. It is not necessarily the ideal or the best possible lighting. It is a **reference** lighting condition only.

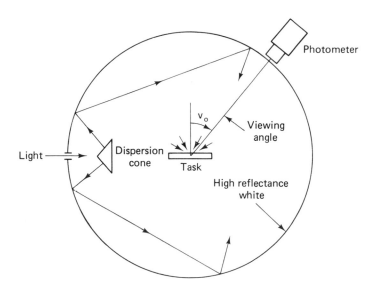

Figure 7-9.　Photometric Sphere

170

It is possible and desirable to have higher ESI footcandles than ordinary footcandles.

100 ESI footcandles—calculated
50 footcandles—measured (classical footcandles)

Classical footcandles are related directly to the watts consumed, so an energy-efficient design involves achieving the highest visibility (ESI fc) with the lowest classical footcandles. Classical footcandles are directly related to energy consumption.

The second key phrase in the example is **the same visibility.** The ESI method **equates** the visibility of a task in the real environment to its visibility under reference lighting conditions (sphere). The question is what level of illumination is required inside the sphere to provide the same visibility as that produced in the real environment. That is, what is the ESI?

Visibility is proportional to one's sensitivity to contrast multiplied by the contrast available.

$$\text{visibility} = \text{RCS} \times C = \text{gain} \times \text{signal}$$

where RCS = relative contrast sensitivity (gain)
C = contrast (signal)

visibility under reference conditions = visibility in the real environment

contrast in sphere	sensitivity to contrast in sphere	=	contrast in real environment	×	sensitivity to contrast in real environment
C_o	× RCS_e	=	C	×	RCS

where C_o = contrast in the sphere; it is calculated or measured and is a constant for a given task
C = contrast in the real environment; it is calculated or measured
RCS = sensitivity to contrast on a relative basis as a function of the task background luminance in the real environment in Table 7-10.

What level of sensitivity is required in the sphere (RCS_e) for the visibility to be the same as that found in the real environment?

$$C_o \times \text{RCS}_e = C \times \text{RCS}$$

Therefore,

$$\text{RCS}_e = \frac{C}{C_o} \times \text{RCS}$$

where

$$\frac{C}{C_O} = \text{CRF (Contrast Rendering Factor)}$$

Once RCS_e or the required level of sensitivity to contrast in the sphere is calculated, L_e, the equivalent background luminance inside the sphere can be determined from the basic visibility curve or Table 7-10. L_e is the level of background luminance required under reference lighting conditions (sphere) to produce the level of contrast sensitive to make the task visibilities the same.

$$RCS_e \rightarrow \text{visibility cure} \rightarrow L_e$$

The diffuse reflectance of the task background in the sphere, ρ_o, can be calculated or measured and is a constant for a given task.

The ESI value is then calculated by

$$\text{ESI} = \frac{L_e}{\rho_O}$$

The mathematics to calculate ESI is quite simple and straightforward once the numbers are known. The calculated ESI can then be compared to a recommended level of equivalent sphere illumination, E_r. The E_r values are established for each type of visual task and can be found in reference 7. E_r is a *recommended* level that should be used as a guideline in evaluating task visibility in a particular environment.

Figure 7-10 gives a flow diagram of the method just described for determining the ESI of a task in a room. The CRF is calculated from visual task photometer (VTP) measurements. The illumination on the task, E_t, and the

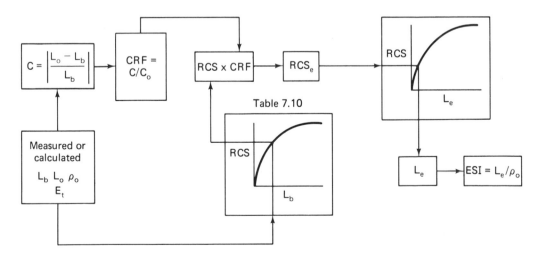

Figure 7-10. Flow Diagram for the Calculation of Equivalent Sphere Illumination

luminance of the task background, L_b, are measured. The L_b value is entered into the RCS curve (or Table 7-10) to find an RCS for the task background luminance. An equivalent RCS (RCS_e) is found by multiplying the CRF by RCS. This takes into account the veiling reflections. The RCS_e is then reentered into the RCS curve (or Table 7-10) and L_e is found. Dividing this equivalent luminance, L_e, by the diffuse reflectance characteristics of the task, ρ_o, gives the equivalent sphere illumination.

A brief examination of the problem makes the calculation of ESI seem quite simple. What makes the method so complex that it requires the use of a computer? The problem is the determination of CRF, specifically the task contrast in the real environment, C, since C_o, the contrast in the sphere, is a constant for a given task.

$$RCS_e = \frac{C}{C_o} \times RCS$$

where

$$\frac{C}{C_o} = \text{CRF (contrast rendition factor)}$$

What is needed to calculate C? The value of C is arrived at by the Blackwell contrast formula:

$$C = \left| \frac{L_o - L_b}{L_b} \right|$$

The factors that affect the contrast in a real environment (C value) are the following:

1. Room dimensions.
2. Room surface reflectances.
3. Luminaire layout.
4. Luminaire candlepower distribution and polarization.
5. Physical characteristics of the task.
6. Viewing angle.
7. Test location.
8. Viewing direction.

Thus it is actually the calculation of the object and background luminance that makes the procedure so complicated. The equations involved in calculating L_b and L_o are

$$L_b = \Sigma [\beta_{(h, v_s)b} \times E_{(h, v_s)t}]$$
$$L_o = \Sigma [\beta_{(h, v_s)o} \times E_{(h, v_s)t}]$$

Since $\beta (h, v_s)_b$ and $\beta (h, v_s)_o$ (bidirectional reflection distribution function; see Chap. 3) are constants for a given task, the problem actually comes down to the calculation of illumination at a point. The illumination at a point is made up of two components: (1) direct and (2) interreflected illumination. As long as the luminaires can be divided into small areas, the direct component calculation simply involves the inverse-square law and the cosine law of illumination. Once the luminances of the room surfaces are determined, each surface can be treated as an emitter. That is, each surface becomes a source of light that can be divided into small areas so that the inverse-square law and the cosine law of illumination can be used to calculate the illumination at a point. This process can be used in determining the contribution from the interreflected component. The calculation of the surface luminances involves the use of radiative luminous flux transfer theory. The computer is necessary to handle the volumes of information, data, and calculations required to determine C.

There are a number of programs commercially available for predetermining VCP and/or ESI. The computational techniques involved are presented in references 5 and 8. Because of the size of the program and its storage requirements, these programs require large computers, which are not economically feasible for most consultants. Therefore, most of these programs are available through time-sharing systems.

Cost Factors

Energy management and energy-efficient design have a tremendous impact on cost. As always, the final decision on which lighting system to install should be based on cost, which should not only include the first cost (initial) of the installation, but also the operating and maintenance costs. With the rise in energy cost and inflation in all sectors of the economy, an inexpensive system (low initial cost) could cost the owner many times more to operate and maintain.

Initial Cost

Energy-efficient design will result in an increase in initial cost over the traditional approach of **"make it cheap."** Cheaper, less efficient light sources and systems will have to be analyzed in terms of total cost. However, overobsession with efficiency can lead to some disasterous effects on color and color psychology.

The statement "better and energy efficient lighting will cost more initially" is not applicable to every situation with certainty. It is probably true, but each situation should be subjected to the economic test. So many factors are involved that the decision on which system to use should not be made on initial cost (either the lowest or highest) alone. A number of examples[9-13] of lighting system economics can be found in the literature.

Maintenance Cost

Maintenance cost is tied directly to initial cost in the selection of lighting equipment. In general, initial cost will increase with more efficient equipment and better maintenance characteristics of the equipment. Durable, well-constructed equipment with fast, easy access will help to minimize maintenance cost. Increased maintenance programs are essential to reduced operating costs. A good maintenance program will minimize light loss due to dirt accumulation and surface deterioration, which is usually compensated for by putting more light into a space. The increase in lighting levels means an increase in the installed watts, which will increase the operating cost.

Operating Cost

Operating cost is tied directly into the amount of power consumed. System design in terms of light source efficiency and overall system efficiency (light source plus auxiliaries plus luminaire performance) will determine the operating cost.

One factor that affects the analysis of the operating alternative is the electrical or utility rate. Whether an average cost per kilowatt hour or the actual cost per kilowatt hour is used greatly affects the outcome of the analysis. Utility rates are based on a number of different schemes, such as block rates (which may have an increasing or decreasing charge with increased usage) and demand rates (which are based on the required capacity of the utility to supply power). As can be seen, the utility rate can be a complex maze of numbers from which the designer has to determine the rate that will be entered into the analysis. It is, therefore, mandatory that the designer accurately determine these costs.

Cost Summary

A comparison of the cost of different lighting systems leaves out many other factors involved in the total building cost. The lighting affects the mechanical system because of heat gain from the luminaires, and heat gain and loss through windows. The floor areas, floor to ceiling heights, and floor-to-floor heights affect building volume, which again affects the mechanical and lighting systems. A discussion[13] of the total building energy picture shows that changes in lighting have little effect on the total energy used, but do greatly affect the productivity of workers.

There are a number of methods of economic analysis by which to review the different choices. The method chosen depends on the purpose and use to which the analysis is put, as well as the factors of interest in the analysis. Four types of analysis are (1) the payback period, (2) the internal rate of return, (3) the present value, and (4) the savings investment ratio. With the recent tremendous increases in the cost of energy, an inflationary factor is critical to the analysis of the operating costs. Operating cost is an ongoing cost and will continue to escalate.

State-of-the-Art in Quality Design

The development of design techniques that evaluate visibility in terms of visual performance has brought about rapid changes in the state of the art in lighting. The impact of the ESI method on lighting technology has been quite important. The designer finally has a handle on one of the major factors that affects how well one can see task detail, and hence perform the task. The designer can now evaluate one of the **quality** aspects of lighting design. With the development of a computer program[14], the designer can predetermine the ESI levels produced in a luminous environment, and truly evaluate a lighting system in terms of visibility. Two ESI instruments [15,16] have been constructed and tested that permits the measurement of ESI values in a space.

With the continuing development of ESI instrumentation, the concern over the inability to measure ESI can seemingly be put aside. As the ESI method gains in acceptance, additional instruments and techniques for quantifying and evaluating ESI will be developed. BRDF's have been measured and are available for at least seven additional tasks. Additional tasks are being studied and will be measured in the very near future. Studies[17] are being conducted on the sensitivity of the ESI system to various changes, such as different layouts, different luminaires, different spacing ratios, and different tasks.

Blackwell has developed refinements to the original CIE report 19[18]. The refinements and expansion of the ESI methodology will appear in a second report of the CIE Committee TC-3.1. The report is entitled "Implementation Procedures for Evaluating Visual Performance Aspects of Lighting." The essence of the report is covered in reference 19. The expanded method described by Blackwell leads to the calculation of relative performance, RP, as a function of reference illumination for different values of $C, \alpha, V, e,$ and VLM, where

$\overline{C}$ = equivalent contrast which is the contrast of the visibility reference task that has the same visibility as the actual task

α = a measure of the difficulty of ocular search and scan, and off-axis information processing

V = proportion of the visual component in a task being performed; if $V = 1$, the task being performed is completely visual

e = error penalty, which is a weighting coefficient for errors

VLM = visibility level multiplier, which accounts for differences in contrast sensitivity for different individuals of the working population; the justification or need for higher or lower levels of reference illumination, E_{ref}, can be demonstrated in terms of the five variables (listed above) that influence task performance

"Values of RP . . . produced by a change in the level of reference illumination should prove very useful in connection with cost benefit studies of lighting applications[19]."

This expanded procedure that expresses visibility in terms of relative performance can be combined with cost benefit analysis[20] to give a more complete indication of trade-offs in terms of energy conservation, productivity, and profits. "You'll not use the new CIE system without a significant increase in the money expended on technological services by the engineers who design the lighting[19]." The design engineer will not only spend additional time, but must have a greater understanding of the new system to evaluate the five variables.

As more data on the relationship of productivity to lighting are gathered and a dollar amount is attached to productivity, the economic analysis will be based on tangible figures rather than a recommended footcandle or ESI level. The relationship of productivity to the quality and quantity of illumination is the subject of current research.

The current computer techniques used for evaluating ESI are analysis procedures. That is, the designer must investigate different combinations of luminaires and layouts to determine which system gives the best visibility for the dollars budgeted to the project. If a criterion level of ESI is specified, the designer must also investigate different design combinations to find a system that will meet that criterion. A true design procedure would involve specifying a given ESI level at preselected task locations and having the computer determine the optimum combination of luminaires and layout to meet that criterion. This type of optimization procedure was first described by Ngai[21,22]. Because of the complexity of optimization, progress in development of the procedure has been slow. There has been considerable discussion of optimization programs, but to date no one has made any information public. The same can be said for snythesis programs. A synthesis program would bring together all aspects of the environment by optimizing the artificial lighting, daylighting, and thermal considerations in terms of cost benefits. This type of "super" program may become a reality in the future, but at this time computer technology (memory size) and man's understanding of the complex interactions are not at that level of sophistication.

references

1. "The Determination of Illumination at a Point in Interior Spaces," Committee on Lighting Design Practice of the IES, LM43, Apr. 1973.

2. Helms, R. N., "An Engineering Approach to the Zonal Cavity Technique," **Illuminating Engineering Journal,** vol. 63, no. 5, May 1968, pp. 287–291.

3. Committee on Recommendations of Quality and Quantity of Illumination of the IES, "Outline of a Standard Procedure for Calculating Visual Comfort Ratings for Interior Lighting—Report No. 2," **Illuminating Engineering,** Oct. 1966, pp. 643–666.

4. Guth, S. K., "Computing Visual Comfort Ratings for a Specific Interior Lighting Installation," **Illuminating Engineering,** Oct. 1966.

5. DiLaura, D. L., "On the Computation of Visual Comfort Probability," **Journal of the Illuminating Engineer Society,** July 1976.

6. Florence, N., "Comparison of the Energy Effectiveness of Office Lighting Systems," paper presented at the 1976 IES Annual Technical Convention, Cleveland, Ohio.

7. Kaufman, John, editor, **IES Lighting Handbook,** 5th ed. Illuminating Engineering Society, New York 1972.

8. DiLaura, D. L., "On the Computation of Equivalent Sphere Illumination," **Journal of the Illuminating Engineering Society,** Jan. 1975.

9. Griffith, J. W., "Response Optimization and Economic Planning," **Lighting Design and Application,** Sept. 1973, pp. 23–27.

10. Mangold, S. A., "Lighting Economics Based on Proper Maintenance," **Lighting Design and Application,** Aug. 1974, pp. 6–11.

11. Finn, J. F., "A Lighting System Cost Comparison," **Lighting Design and Application,** Jan. 1975, pp. 26–27.

12. Lange, A. W., "New Design and Specification Techniques Cut Energy Costs and Offer Optimum Efficiency," **Lighting Design and Application,** Feb. 1976, pp. 10–13.

13. Dorsey, R. T., "Cost–Benefit Analysis Applied to Lighting in the Energy Equation," **Lighting Design and Application,** July 1975, pp. 36–38.

14. Lumen II, a program developed by Dave L. DiLaura; Smith, Hinchman, and Grylls, Detroit, Mich.

15. Ngai, P. Y., Zeller, R. D., and Griffith, J. W., "The ESI Meter—Theory and Practical Embodiment," **Journal of the Illuminating Engineering Society,** Oct. 1975, pp. 58–65.

16. DiLaura, D. L., and Stannard, S. M., "An Instrument for the Measurement of Equivalent Sphere Illumination," **Journal of the Illuminating Engineering Society,** Apr. 1978, pp. 183–187.

17. Lewin, I., "An ESI Study for Different Tasks," a paper presented at the 1976 IES Annual Technical Convention, Cleveland, Ohio.

18. "Recommended Method for Evaluating Visual Performance Aspects of Lighting," CIE Report 19, prepared by Committee E-1.4.2 on Visual Performance, available through Mr. Jack Tech, Secretary USNC–CIE, National Bureau of Standards, Washington, D.C., 20234.

19. Blackwell, H. R., "Energy Conservation by Selective Lighting Standards Graded in Terms of Task and Observer Characteristics," Proceedings—The Basis for Effective Management of Lighting Energy Symposium, Federal Energy Administration, Washington, D.C., Oct. 29, 30, 1975.

20. Dorsey, D. T., and Blackwell, R. H., "A Performance-Oriented Approach to Lighting Specifications," **Lighting Design and Application,** Feb. 1975, pp. 13–27.

21. Ngai, P. Y., "Veiling Reflections and the Design of the Optimal Intensity Distribution of a Luminaire in Terms of Visual Performance Potential," **Journal of the Illuminating Engineering Society,** Oct. 1974, pp. 53–59.

22. Ngai, P. Y., and Helms, R. N., "Optimization—A Synthetic Approach to Lighting Design," **Journal of the Illuminating Engineering Society,** July 1975, pp. 286–291.

23. Helms, R. N., **Illuminating Engineering Workbook,** 2nd Ed., University of Colorado Publication Service, Boulder, Colorado, 1979.

data section

(For additional data see reference 23)

TABLE 7-1

Photometric Test Data

Description:

Incandescent wall wash unit
w/150PAR38/SP rated at 1750
lumens.

Report No. 10500

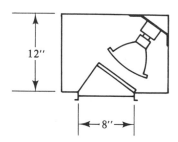

Candlepower

Angle	Horz*	Vert+
0	10000	1020
5	9730	2380
10	9300	4520
15	8700	6530
20	7910	8150
25	6790	9430
30	3150	10000
35	2400	9100
40	910	7360
45	105	5400
50	10	3180
55	0	1430
60	0	100
65	0	15
70	0	0
75	0	0
80	0	0
85	0	0
90	0	0

*Horz — Horizontal Plane
through the aiming point.

+Vert — Vertical Plane
aiming at 30°

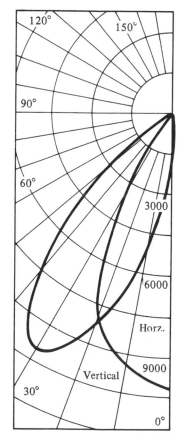

179

TABLE 7-2

Typical Fluorescent Coefficient of Utilization Table

Lumen method problem
Luminaire No. DB117
(w/4F40WW at 3100 lm each)
Zonal cavity coefficients of utilization: eff. floor cav. ref. = 0.20

Ceiling wall RCR	80			70			50			10			0
	50	30	10	50	30	10	50	30	10	50	30	10	0
1	0.68	0.66	0.64	0.67	0.65	0.63	0.64	0.63	0.61	0.59	0.58	0.57	0.56
2	0.62	0.58	0.55	0.60	0.57	0.54	0.58	0.56	0.53	0.54	0.52	0.51	0.49
3	0.55	0.51	0.48	0.54	0.51	0.48	0.53	0.49	0.46	0.50	0.47	0.45	0.44
4	0.50	0.45	0.42	0.49	0.45	0.41	0.48	0.44	0.41	0.45	0.42	0.40	0.39
5	0.46	0.40	0.36	0.44	0.40	0.36	0.43	0.39	0.36	0.41	0.38	0.35	0.34
6	0.41	0.36	0.32	0.40	0.35	0.32	0.39	0.35	0.32	0.37	0.34	0.31	0.30
7	0.37	0.32	0.28	0.37	0.31	0.28	0.35	0.31	0.28	0.34	0.30	0.27	0.26
8	0.33	0.28	0.25	0.33	0.28	0.24	0.32	0.27	0.24	0.30	0.27	0.24	0.23
9	0.30	0.25	0.21	0.29	0.24	0.21	0.29	0.24	0.21	0.27	0.24	0.21	0.20
10	0.27	0.22	0.19	0.27	0.22	0.19	0.26	0.22	0.19	0.25	0.21	0.19	0.17

Max S/MH = 1.43

LDD Category V

TABLE 7-3

Typical Fluorescent Photometric Test Report

PHOTOMETRIC DATA

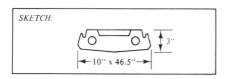

SKETCH:

3"
10" x 46.5"

Description:

1 x 4 surface mounted clear
prismatic acrylic wraparound
w/metal ends

Test No. 11417		*Cat. No.* BK 240	
Lamps 2F40 CW		*Volts* 120	
Lumens/Lamp 3110		*Test Dist.* 20	
Shielding Angle: N 90° L 90°			
Bare Lamp 2360 fL		*Date*	

LDD CATEGORY V

Brightness Data

	Angle	Maximum	Average	Ratio Max./Min.
Along (0°)	45°	2483	1148	2.2
	55°	1939	909	2.1
	65°	1089	678	1.6
	75°	850	533	1.6
	85°	510	384	1.3
Across (90°)	45°	2721	1076	2.5
	55°	2211	827	2.7
	65°	1701	628	2.7
	75°	1429	560	2.6
	85°	1463	597	2.5

Distribution Data

Angle	C.P.	Angle	C.P.
0°	1398	95°	106
5°	1389	105°	116
10°	—	115°	116
15°	1353	125°	115
20°	—	135°	96
25°	1285	145°	75
30°	·	155°	63
35°	1159	165°	49
40°	—	175°	39
45°	937	180°	40
50°	—		
55°	615		
60°	··		
65°	365		
70°	··		
75°	221		
80°	—		
85°	134		
90°	107		

OUTPUT DATA

Zone Degrees	Lumens	Percent Total Lamp Lumens	Percent Total Distance
0–40°	1838	29.5	41.0
40–90°	2019	32.5	45.1
90–180°	625	10.1	13.9
0–180°	4482	72.1	100.0

COEFFICIENT OF UTILIZATION

Floor		ρ_{cc} = 20%								
Ceiling		80%			50%			10%		
Walls		50%	30%	10%	50%	30%	10%	50%	30%	10%
Room Cavity Ratio	1	73	71	68	66	46	62	57	56	55
	2	65	60	56	58	55	52	51	49	47
	3	57	52	48	52	48	45	46	43	41
	4	51	45	41	47	42	39	41	39	35
	5	46	39	35	42	37	33	37	33	31
	6	41	35	31	37	33	29	33	30	27
	7	37	31	27	34	29	25	30	26	24
	8	33	27	23	30	25	22	27	23	21
	9	30	24	20	27	22	19	24	21	28
	10	27	21	18	25	20	17	22	18	16

Maximum recommended spacing-to-mounting height ratio above work plane is: 1.4

TABLE 7-4

Effective Ceiling or Floor Cavity Reflectance

	Effective Cavity Reflectance				

BASE REFLECTANCE PER CENT → 90, 80, 70, 60, 50
WALL REFLECTANCE PER CENT → 90 80 70 60 50 40 30 20 10 0 (for each base group)

ROOM CAVITY RATIO	90-90	90-80	90-70	90-60	90-50	90-40	90-30	90-20	90-10	90-0	80-90	80-80	80-70	80-60	80-50	80-40	80-30	80-20	80-10	80-0	70-90	70-80	70-70	70-60	70-50	70-40	70-30	70-20	70-10	70-0	60-90	60-80	60-70	60-60	60-50	60-40	60-30	60-20	60-10	60-0	50-90	50-80	50-70	50-60	50-50	50-40	50-30	50-20	50-10	50-0
0.2	89	88	88	87	87	86	86	85	85	84	79	78	78	77	77	76	76	75	74	72	70	69	68	68	67	67	66	66	65	64	60	59	59	58	58	57	56	56	55	53	50	50	49	49	48	48	47	46	46	44
0.4	88	87	86	85	84	84	82	81	80	79	79	77	76	75	74	73	71	70	69	66	69	68	67	66	65	64	63	61	59	57	59	59	58	57	56	55	53	51	50	46	50	49	48	48	47	46	45	44	42	41
0.6	87	86	84	83	81	80	77	76	74	73	78	76	73	72	71	70	67	65	64	62	68	67	65	64	63	61	59	58	57	54	58	57	56	55	53	51	50	47	46	41	50	48	47	46	45	44	43	42	41	38
0.8	87	85	82	81	80	79	77	74	71	69	67	78	75	73	71	69	67	65	53	50	48	68	66	64	62	60	58	54	50	48	47	44	58	57	55	53	51	47	44	41	38	50	48	47	45	44	42	38	35	33
1.0	86	83	80	77	75	72	69	66	64	62	77	74	72	69	67	65	62	60	57	52	68	65	62	60	58	55	53	49	45	41	57	55	53	51	49	45	40	37	34	29	50	48	46	44	43	41	38	36	34	34
1.2	85	82	78	75	72	69	66	63	60	57	76	72	69	66	64	61	59	57	53	51	67	64	61	59	57	54	51	48	44	38	59	56	54	51	49	47	44	41	39	36	47	45	43	41	39	36	33	29	27	24
1.4	85	80	77	73	69	65	62	59	57	53	77	73	69	65	60	58	55	51	47	44	63	60	58	55	51	47	45	41	39	35	56	53	49	46	44	41	39	37	35	32	45	43	40	38	35	31	27	25	23	21
1.6	84	79	75	71	67	63	59	56	53	50	75	71	66	63	59	56	53	50	47	44	62	59	56	54	51	46	42	40	38	35	56	53	48	45	43	39	36	33	30	26	44	41	39	36	33	30	26	23	21	20
1.8	83	78	73	69	64	60	56	53	50	47	75	70	66	62	58	54	51	46	42	40	61	58	54	50	46	43	38	35	33	31	54	50	47	44	40	37	34	30	28	25	43	40	38	34	31	28	25	22	20	19
2.0	83	77	72	67	62	57	53	50	47	43	74	69	64	60	56	52	49	45	41	38	66	56	52	49	45	40	38	36	33	29	54	50	46	43	39	36	33	29	26	24	40	37	34	30	28	26	24	21	19	17
2.2	82	76	70	65	59	54	50	47	44	40	74	68	63	58	54	49	45	42	38	35	66	60	55	51	48	43	38	36	34	32	53	49	45	42	37	34	31	29	27	24	38	36	33	29	27	25	23	21		
2.4	82	75	69	64	58	53	48	45	41	37	73	67	61	56	52	47	43	40	36	33	65	60	54	50	46	41	37	35	32	30	53	48	44	41	36	32	30	27	25	23	37	35	31	27	25	23	21			
2.6	81	74	67	62	56	51	46	42	38	35	72	66	60	55	50	45	41	38	34	31	65	59	54	50	45	40	35	33	30	28	48	43	39	35	34	31	28	26	23	21	36	34	30	26	23	21	20			
2.8	81	73	66	60	54	49	44	40	36	34	73	65	59	53	48	43	39	36	32	29	64	58	54	48	43	38	33	30	28	26	47	43	38	34	29	27	24	22			33	31	28	25	22	19	17			
3.0	80	72	64	58	52	47	42	38	34	30	72	65	58	52	47	42	37	34	30	27	64	58	52	47	42	37	32	29	27	24	46	42	37	32	28	24	23	20			32	30	26	23	21	20	17			
3.2	79	71	63	56	50	45	40	36	32	28	72	65	51	45	40	35	33	28	24	18	58	51	46	40	36	31	27	23	22	18	39	35	31	27	23	20	18	16			31	30	27	23	20	18	16			
3.4	79	70	62	54	48	43	38	34	30	27	71	64	56	49	44	39	35	32	27	24	57	51	45	40	35	30	26	22	20	17	35	30	26	23	20	17	15	12			30	26	22	19	17	15				
3.6	78	69	61	53	47	42	36	32	28	25	71	63	54	48	43	38	33	28	25	22	56	49	43	38	33	28	25	22	19	16	34	29	25	21	18	16	14				29	25	21	18	16	14				
3.8	78	69	60	51	45	40	35	31	27	23	70	62	53	47	41	36	32	27	24	19	56	49	43	37	32	27	24	20	17	15	33	29	24	21	17	15	13				29	25	21	17	15	13				
4.0	77	69	58	51	44	39	33	29	25	22	70	61	53	46	40	35	30	26	22	20	55	48	42	36	31	26	23	20	17	14	32	28	23	20	17	14	12				28	24	20	17	15	12				
4.2	77	62	57	50	43	37	32	28	24	21	69	60	52	45	39	34	29	25	21	18	55	47	41	35	30	25	22	19	16	14	32	27	32	19	17	14	12				28	24	20	17	14	12				
4.4	76	61	56	49	42	36	31	27	23	20	69	60	51	44	38	32	27	22	19	15	54	46	40	34	29	24	21	18	15	11	31	27	23	19	16	13	11				27	23	19	16	13	11				
4.6	76	60	55	47	40	35	30	26	22	19	69	59	50	43	37	32	27	23	19	15	53	45	39	33	28	24	20	17	14	12	30	26	22	18	15	13	10				26	22	18	15	13	10				
4.8	75	59	54	46	39	34	28	25	21	18	68	58	49	42	36	31	26	22	18	14	52	44	38	32	27	23	20	16	13	09	31	26	20	16	12	10	07				26	22	18	15	12	09				
5.0	75	59	53	45	38	33	28	24	20	16	68	58	48	41	35	30	26	23	18	14	52	44	38	31	26	22	19	15	12	09	30	25	30	17	14	11	09				25	21	17	14	12	09				
6.0	73	61	49	41	34	29	24	20	16	11	66	51	44	38	31	27	22	17	14	10	51	41	35	28	24	19	16	13	10	06	25	21	17	14	11	07					23	19	15	13	10	06				
7.0	70	58	45	38	31	27	23	18	14	08	64	53	41	35	28	24	19	16	12	07	48	38	32	26	22	17	14	11	08	05	24	20	16	12	09	05					21	18	14	11	08	05				
8.0	68	55	42	35	27	23	18	15	12	06	62	50	38	32	25	21	17	14	11	05	46	35	29	23	18	15	13	10	05	04	22	18	14	11	08	04					19	16	12	10	07	03				
9.0	66	52	38	31	25	21	16	14	11	05	61	49	36	30	23	19	15	13	09	04	45	33	27	21	18	14	11	09	07	03	20	16	12	09	07	03					18	15	11	09	07	03				
10.0	65	51	36	29	22	19	15	11	09	04	59	46	33	27	21	18	14	11	08	03	43	31	25	19	16	12	10	08	06	02	17	14	11	09	07	02					17	14	10	08	06	02				

182

Table 7.4 (cont.)

BASE REFLECTANCE PER CENT / WALL REFLECTANCE PER CENT

ROOM CAVITY RATIO	40										30										20										10										0									
	90	80	70	60	50	40	30	20	10	0	90	80	70	60	50	40	30	20	10	0	90	80	70	60	50	40	30	20	10	0	90	80	70	60	50	40	30	20	10	0	90	80	70	60	50	40	30	20	10	0
0.2	40	40	39	39	39	38	38	37	36	36	31	31	30	30	29	29	28	28	27	27	21	20	20	20	19	19	19	19	18	17	11	11	11	11	10	10	10	10	09	09	02	02	01	01	01	01	01	01	00	00
0.4	41	40	39	39	38	37	36	35	34	34	31	31	30	30	29	28	27	26	25	24	22	21	21	20	19	18	17	17	16	14	12	12	11	11	10	09	09	08	08	07	04	03	03	02	02	02	01	01	01	00
0.6	41	40	39	38	37	36	34	33	32	31	32	31	30	29	28	27	26	24	23	22	23	22	21	20	19	18	17	16	15	13	13	12	11	11	10	09	08	08	07	06	05	05	04	03	03	02	02	02	01	00
0.8	41	40	38	37	36	35	33	32	31	29	32	31	30	29	28	26	25	23	22	20	24	22	21	20	19	18	16	15	14	13	15	13	12	11	10	09	08	07	06	06	07	06	05	04	03	03	02	02	01	00
1.0	42	40	38	37	35	33	32	31	29	27	33	32	30	28	27	25	23	22	20	19	25	23	21	20	18	17	15	14	13	11	16	14	13	11	10	09	08	07	06	05	08	07	06	05	04	03	02	02	01	00
1.2	42	40	38	36	34	32	30	29	27	25	33	32	30	28	26	24	22	20	19	17	25	23	21	19	18	16	14	13	12	10	17	14	12	11	10	08	07	06	05	05	10	08	07	06	04	03	03	02	01	00
1.4	42	39	37	35	33	31	29	27	25	23	34	32	29	27	25	23	21	19	17	16	26	23	21	19	17	15	14	12	11	09	16	14	12	11	09	08	07	06	05	04	11	09	08	06	05	04	03	02	01	00
1.6	42	39	37	35	32	30	27	25	23	22	34	31	29	27	24	22	20	18	16	14	26	23	21	18	16	15	13	11	10	09	17	13	12	10	09	08	06	05	05	04	12	10	08	07	05	04	03	02	01	00
1.8	42	39	36	34	31	29	26	24	22	21	35	31	29	26	23	21	18	16	15	13	27	23	20	18	16	14	12	11	09	08	19	13	11	10	08	07	06	05	04	03	13	11	09	07	06	04	03	02	01	00
2.0	42	39	36	34	31	28	25	23	21	19	35	31	28	26	23	20	18	16	14	12	28	23	20	18	15	14	12	10	09	07	19	13	11	10	08	07	06	05	04	03	14	11	09	07	06	05	03	02	01	00
2.2	42	39	36	33	30	27	24	22	19	18	36	31	28	25	22	20	17	15	13	11	28	23	20	17	15	13	11	09	08	06	21	13	11	09	08	06	05	04	04	03	15	11	09	07	06	05	03	02	01	00
2.4	43	39	35	33	29	27	24	21	18	17	36	31	27	24	21	19	16	14	12	10	29	23	20	17	14	12	10	09	07	06	22	13	11	09	07	06	05	04	03	02	16	11	09	07	06	05	03	02	01	00
2.6	43	39	35	32	29	26	23	20	17	15	36	31	27	24	21	18	16	13	11	09	29	23	19	16	14	12	10	08	07	05	23	13	10	09	07	06	05	04	03	02	17	12	09	07	06	05	03	02	01	00
2.8	43	39	35	32	28	25	22	19	16	14	37	31	27	24	20	18	15	13	10	08	30	23	19	16	13	11	09	08	06	05	23	13	10	08	07	06	04	03	03	02	17	12	09	07	06	05	03	02	01	00
3.0	43	39	35	31	27	24	21	18	16	13	37	31	27	23	20	17	14	12	10	08	30	23	19	16	13	11	09	07	06	04	24	13	10	08	07	05	04	03	02	02	18	12	09	07	05	05	03	02	01	00
3.2	43	39	35	31	27	23	20	17	15	13	37	31	26	23	19	16	14	11	09	07	31	23	19	15	13	10	08	07	05	04	25	13	10	08	06	05	04	03	02	02	19	12	09	07	05	04	03	02	01	00
3.4	43	39	34	30	26	23	20	17	14	12	37	31	26	22	19	16	13	11	09	07	31	23	18	15	12	10	08	06	05	03	26	13	10	08	06	05	04	03	02	01	20	13	09	07	05	04	03	02	01	00
3.6	44	39	34	30	26	22	19	16	14	11	38	31	26	22	18	15	13	10	08	06	32	23	18	15	12	10	08	06	05	03	26	13	09	07	06	05	04	03	02	01	20	13	10	08	05	04	03	02	01	00
3.8	44	38	33	29	25	22	18	16	13	10	38	30	25	21	18	15	12	10	08	05	32	22	18	14	11	09	07	05	04	02	27	13	09	07	06	05	04	03	02	01	21	13	10	08	05	04	03	02	01	00
4.0	44	38	33	29	25	21	18	15	12	10	38	30	25	21	17	14	12	09	08	05	33	22	17	14	11	09	07	05	04	02	27	14	09	07	06	04	03	02	02	01	22	14	10	08	06	04	03	02	01	00
4.2	44	38	33	29	24	21	17	15	12	10	38	30	25	21	17	14	11	09	07	04	33	22	17	14	11	09	07	05	04	02	28	14	09	07	06	04	03	02	02	01	22	14	10	08	06	04	03	02	01	00
4.4	44	38	33	28	24	20	16	14	11	09	39	30	24	20	16	14	11	08	06	04	34	22	17	13	10	08	06	04	03	02	28	13	09	07	06	04	03	02	02	01	23	14	10	08	06	04	03	02	01	00
4.6	44	38	32	28	23	19	16	14	11	08	39	30	24	20	16	13	10	08	06	03	34	22	17	13	10	08	06	04	03	01	29	13	09	07	05	04	03	02	01	01	23	14	10	08	06	04	03	02	01	00
4.8	44	38	32	27	22	19	16	13	10	08	39	30	24	19	15	13	10	08	05	03	35	22	17	13	10	08	06	04	03	01	29	13	08	06	05	04	03	02	01	01	24	14	10	08	06	04	03	02	01	00
5.0	45	38	31	27	22	19	15	13	10	07	40	30	24	19	15	12	09	07	05	02	35	22	16	13	09	08	05	04	03	01	30	14	08	06	05	04	03	02	01	01	25	14	11	08	06	04	03	02	01	00
6.0	44	37	30	25	20	17	13	11	08	05	39	29	22	17	13	10	08	06	03	01	36	20	15	11	08	06	04	03	02	01	31	11	07	05	04	03	02	01	01	01	27	15	11	08	06	05	03	02	01	00
7.0	44	36	29	24	19	16	12	10	07	04	40	28	21	16	12	09	07	04	03	01	36	20	14	10	07	05	04	03	02	01	32	11	07	05	04	03	02	01	01	01	28	15	11	08	06	04	03	02	01	00
8.0	44	35	28	23	18	15	11	09	06	03	40	28	20	15	11	08	06	04	02	01	37	19	13	09	07	05	03	02	01	01	33	11	06	04	03	02	02	01	01	01	30	15	12	08	06	04	03	02	01	00
9.0	44	35	26	21	16	13	10	08	05	02	40	28	19	14	10	08	05	04	02	01	37	18	12	08	06	04	03	02	01	01	34	10	06	04	03	02	01	01	01	01	31	15	12	09	06	04	03	02	01	00
10.0	43	34	25	20	15	12	08	07	05	02	40	27	18	13	10	07	05	03	02	01	37	18	12	08	06	04	03	02	01	01	34	10	06	04	03	02	01	01	01	01	31	15	12	09	06	04	03	02	01	00

183

TABLE 7-5

Effective Floor Cavities (Courtesy of the Illuminating Engineering Society of North America)

For 30 per cent effective floor cavity reflectance, multiply by appropriate factor below.
For 10 per cent effective floor cavity reflectance, divide by appropriate factor below.

Per Cent Effective Ceiling Cavity Reflectance, ρ_{CC}	80			70			50			10		
Per Cent Wall Reflectance, ρ_W	50	30	10	50	30	10	50	30	10	50	30	10
Room Cavity Ratio												
1	1.08	1.08	1.07	1.07	1.06	1.06	1.05	1.04	1.04	1.01	1.01	1.01
2	1.07	1.06	1.05	1.06	1.05	1.04	1.04	1.03	1.03	1.01	1.01	1.01
3	1.05	1.04	1.03	1.05	1.04	1.03	1.03	1.03	1.02	1.01	1.01	1.01
4	1.05	1.03	1.02	1.04	1.03	1.02	1.03	1.02	1.02	1.01	1.01	1.00
5	1.04	1.03	1.02	1.03	1.02	1.02	1.02	1.02	1.01	1.01	1.01	1.00
6	1.03	1.03	1.01	1.03	1.02	1.01	1.02	1.02	1.01	1.01	1.01	1.00
7	1.03	1.02	1.01	1.03	1.02	1.01	1.02	1.01	1.01	1.01	1.01	1.00
8	1.03	1.02	1.01	1.02	1.02	1.01	1.02	1.01	1.01	1.01	1.01	1.00
9	1.02	1.01	1.01	1.02	1.01	1.01	1.02	1.01	1.01	1.01	1.01	1.00
10	1.02	1.01	1.01	1.02	1.01	1.01	1.02	1.01	1.01	1.01	1.01	1.00

TABLE 7-6 (cont.)

LDD Graphs

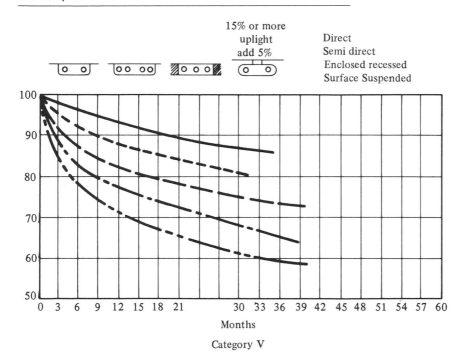

Category V

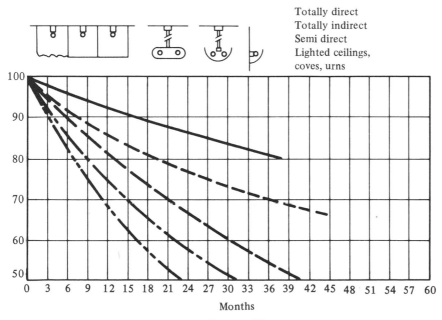

Category VI

TABLE 7-7

Lamp Lumen Depreciation (LDD) 70 percent of life.
(Courtesy of the Illuminating Engineering Society of
North America)

Lamp Description		LLD Factor
Incandescent		
General service	to 150 W	91
	250 to 500 W	90
	750 to 1500 W	86
Silver-bowl	200 to 500 W	75
Reflector	R40	86
	R52 and R57	81
Projector	PAR 38 to 64	84
Mercury		
H39-22 KB	175 W	85
H39-22 KC/C	175 W	83
H39-22 KC/W	175 W	75
H37-5 KB	250 W	85
H37-5 KC/C	250 W	83
H37-5 KC/W	250 W	73
H33-1 CD	400 W	86
H33-1 GL/C	400 W	83
H33-1 GL/W	400 W	74
H36-15 GV	1000 W	77
H36-15 GW/C	1000 W	72
H36-15 GW/W	1000 W	61

	Hours per Start		
Fluorescent	6	12	18
Instant start 425 ma			
Standard colors	88	87	85
Improved-color types	82	80	78
Rapid start 430 ma			
Standard colors	87	86	85
Improved-color types	81	80	79
Rapid start 800 ma			
Standard colors	81	79	77
Rapid start 1500 ma			
Tubular	76	74	72
Others	70	68	64

TABLE 7-8

Spacing Requirements (Courtesy of the Illuminating Engineering Society of North America)

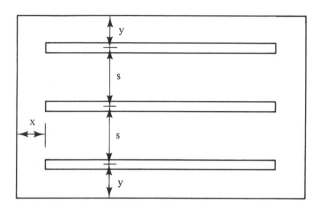

Recommended spacing for "uniformity"

Distance	Preferred	Maximum
x	6" to 12"	2'-0"
y	$2\frac{1}{2}'$ to 3'	S/2

(a)

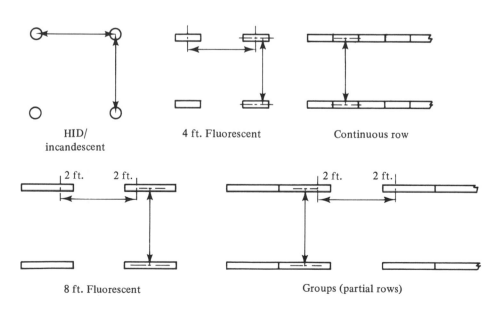

HID/
incandescent

4 ft. Fluorescent

Continuous row

8 ft. Fluorescent

Groups (partial rows)

(b)

TABLE 7-9

Typical VCP Table

Description: 2 × 4 troffer w/6250 lens w/4F40CW, each rated at 3100 lm					LUMNO = DB117 Report No. 1857-1			
100 fc room		Reflectances 80/50/20 Luminaires lengthwise				Luminaires crosswise		
W	L	8.5	10.0	13.0	16.0	8.5	10.0	13.0	16.0
20	20	71	74	76	77	70	73	75	75
20	30	64	68	72	73	63	67	70	71
20	40	61	64	68	69	60	63	66	67
20	60	58	60	64	66	56	59	62	64
30	20	71	75	77	76	71	75	76	75
30	30	64	68	72	72	63	67	70	70
30	40	59	63	67	68	59	62	65	66
30	60	56	59	62	64	55	58	61	62
30	80	53	57	59	61	52	55	58	59
40	20	72	76	79	78	72	76	78	76
40	30	64	68	73	73	64	68	72	71
40	40	59	63	67	69	59	62	66	67
40	60	56	58	61	64	55	57	60	62
40	80	53	56	58	60	52	55	57	59
40	100	52	54	56	58	51	53	55	57
60	30	65	69	73	74	65	68	72	73
60	40	60	63	66	69	59	62	66	68
60	60	56	58	60	63	55	57	60	62
60	80	53	55	57	60	52	54	56	58
60	100	52	54	55	57	51	53	54	56
100	40	62	65	68	71	62	65	67	70
100	60	58	60	61	64	57	59	61	63
100	80	55	57	57	60	54	56	57	59
100	100	53	54	55	57	52	54	54	56

190

TABLE 7-10

Relative Contrast Sensitivity, RCS (Courtesy of the Illuminating Engineering Society of North America)

RCS	L	RCS	L	RCS	L	RCS	L	RCS	L
100.	2,920.	93.7	944.	87.4	461.	81.1	229.	74.8	115.9
99.9	2,767.	93.6	930.	87.3	456.	81.0	226.	74.7	114.7
99.8	2,606.	93.5	920.	87.2	452.	80.9	224.	74.6	113.5
99.7	2,550.	93.4	910.	87.1	447.	80.8	222.	74.5	112.3
99.6	2,458.	93.3	900.	87.0	442.	80.7	220.	74.4	111.0
99.5	2,390.	93.2	888.	86.9	437.	80.6	218.	74.3	109.8
99.4	2,327.	93.1	878.	86.8	433.	80.5	215.	74.2	108.4
99.3	2,264.	93.0	865.	86.7	428.	80.4	213.	74.1	107.2
99.2	2,206.	92.9	856.	86.6	423.	80.3	210.	74.0	106.0
99.1	2,153.	92.8	844.	86.5	419.	80.2	208.	73.9	105.0
99.0	2,110.	92.7	834.	86.4	414.	80.1	206.	73.8	103.9
98.9	2,062.	92.6	824.	86.3	410.	80.0	204.	73.7	102.5
98.8	2,022.	92.5	815.	86.2	405.	79.9	201.	73.6	101.5
98.7	1,979.	92.4	805.	86.1	401.	79.8	199.0	73.5	100.2
98.6	1,938.	92.3	795.	86.0	397.	79.7	196.3	73.4	99.0
98.5	1,900.	92.2	785.	85.9	392.	79.6	194.2	73.3	97.8
98.4	1,868.	92.1	776.	85.8	388.	79.5	192.2	73.2	96.6
98.3	1,835.	92.0	767.	85.7	384.	79.4	190.0	73.1	95.4
98.2	1,810.	91.9	758.	85.6	379.	79.3	188.0	73.0	94.2
98.1	1,780.	91.8	749.	85.5	375.	79.2	186.0	72.9	93.0
98.0	1,750.	91.7	741.	85.4	371.	79.1	183.5	72.8	92.0
97.9	1,721.	91.6	732.	85.3	367.	79.0	181.5	72.7	91.1
97.8	1,692.	91.5	724.	85.2	363.	78.9	179.5	72.6	90.0
97.7	1,664.	91.4	717.	85.1	358.	78.8	177.5	72.5	88.9
97.6	1,635.	91.3	708.	85.0	354.	78.7	175.5	72.4	87.8
97.5	1,608.	91.2	700.	84.9	350.	78.6	173.6	72.3	86.7
97.4	1,581.	91.1	692.	84.8	346.	78.5	171.5	72.2	85.8
97.3	1,559.	91.0	682.	84.7	342.	78.4	169.8	72.1	84.8
97.2	1,534.	90.9	676.	84.6	339.	78.3	167.8	72.0	83.7
97.1	1,510.	90.8	670.	84.5	335.	78.2	166.1	71.9	82.8

191

TABLE 7-10 (cont.)

RCS	L	RCS	L	RCS	L	RCS	L	RCS	L
97.0	1,484.	90.7	662.	84.4	330.	78.1	164.6	71.8	82.2
96.9	1,460.	90.6	655.	84.3	327.	78.0	162.9	71.7	81.3
96.8	1,440.	90.5	648.	84.2	323.	77.9	161.0	71.6	80.5
96.7	1,420.	90.4	640.	84.1	320.	77.8	159.0	71.5	79.5
96.6	1,400.	90.3	634.	84.0	316.	77.7	157.3	71.4	78.7
96.5	1,380.	90.2	626.	83.9	312.	77.6	155.9	71.3	77.8
96.4	1,363.	90.1	620.	83.8	309.	77.5	154.1	71.2	77.0
96.3	1,340.	90.0	614.	83.7	306.	77.4	152.4	71.1	76.2
96.2	1,320.	89.9	608.	83.6	302.	77.3	151.0	71.0	75.4
96.1	1,300.	89.8	602.	83.5	299.	77.2	149.5	70.9	74.6
96.0	1,280.	89.7	596.	83.4	295.	77.1	148.0	70.8	74.0
95.9	1,262.	89.6	590.	83.3	292.	77.0	146.5	70.7	73.2
95.8	1,245.	89.5	584.	83.2	289.	76.9	145.1	70.6	72.4
95.7	1,230.	89.4	577.	83.1	286.	76.8	143.6	70.5	71.8
95.6	1,212.	89.3	571.	83.0	283.	76.7	142.1	70.4	71.1
95.5	1,196.	89.2	565.	82.9	280.	76.6	140.7	70.3	70.4
95.4	1,179.	89.1	559.	82.8	276.	76.5	139.3	70.2	69.8
95.3	1,165.	89.0	552.	82.7	274.	76.4	137.8	70.1	69.1
95.2	1,149.	88.9	546.	82.6	271.	76.3	136.4	70.0	68.4
95.1	1,134.	88.8	540.	82.5	268.	76.2	134.9	69.9	67.7
95.0	1,118.	88.7	534.	82.4	265.	76.1	133.4	69.8	67.0
94.9	1,105.	88.6	528.	82.3	262.	76.0	132.0	69.7	66.2
94.8	1,091.	88.5	522.	82.2	259.	75.9	130.5	69.6	65.6
94.7	1,078.	88.4	516.	82.1	256.	75.8	129.0	69.5	65.0
94.6	1,063.	88.3	510.	82.0	253.	75.7	127.5	69.4	64.3
94.5	1,048.	88.2	504.	81.9	250.	75.6	126.1	69.3	63.7
94.4	1,034.	88.1	500.	81.8	248.	75.5	125.0	69.2	63.0
94.3	1,019.	88.0	493.	81.7	245.	75.4	123.5	69.1	62.4
94.2	1,007.	87.9	487.	81.6	242.	75.3	122.3	69.0	61.8
94.1	995.	87.8	482.	81.5	240.	75.2	120.9	68.9	61.1
94.0	982.	87.7	477.	81.4	239.	75.1	119.6	68.8	60.4
93.9	970.	87.6	471.	81.3	235.	75.0	118.5	68.7	59.8
93.8	956.	87.5	467.	81.2	232.	74.9	117.0	68.6	59.2

TABLE 7-10 (cont.)

RCS	L		RCS	L		RCS	L		RCS	L		RCS	L
68.5	58.5		62.0	28.5		55.5	14.88		49.0	8.23		42.6	4.80
68.4	57.9		61.9	28.2		55.4	14.73		48.9	8.17		42.5	4.76
68.3	57.2		61.8	28.0		55.3	14.58		48.8	8.11		42.4	4.72
68.2	56.6		61.7	27.7		55.2	14.42		48.7	8.03		42.3	4.68
68.1	56.0		61.6	27.4		55.1	14.27		48.6	7.96		42.2	4.64
68.0	55.4		61.5	27.1		55.0	14.15		48.5	7.90		42.1	4.60
67.9	54.8		61.4	26.8		54.9	14.01		48.4	7.83		42.0	4.56
67.8	54.1		61.3	26.6		54.8	13.86		48.3	7.77		41.9	4.52
67.7	53.5		61.2	26.3		54.7	13.72		48.2	7.71		41.8	4.48
67.6	52.9		61.1	26.0		54.6	13.60		48.1	7.65		41.7	4.44
67.5	52.3		61.0	25.8		54.5	13.45		48.0	7.59		41.6	4.40
67.4	51.6		60.9	25.6		54.4	13.32		47.9	7.53		41.5	4.36
67.3	51.0		60.8	25.3		54.3	13.19		47.8	7.47		41.4	4.32
67.2	50.4		60.7	25.0		54.2	13.08		47.7	7.41		41.3	4.29
67.1	50.0		60.6	24.79		54.1	12.93		47.6	7.35		41.2	4.25
67.0	49.3		60.5	24.54		54.0	12.81		47.5	7.29		41.1	4.21
66.9	48.7		60.4	24.32		53.9	12.69		47.4	7.23		41.0	4.17
66.8	48.1		60.3	24.04		53.8	12.58		47.3	7.17		40.9	4.14
66.7	47.6		60.2	23.82		53.7	12.43		47.2	7.12		40.8	4.11
66.6	47.0		60.1	23.53		53.6	12.31		47.1	7.05		40.7	4.08
66.5	46.4		60.0	23.32		53.5	12.18		47.0	6.99		40.6	4.04
66.4	45.8		59.9	23.08		53.4	12.08		46.9	6.93		40.5	4.01
66.3	45.3		59.8	22.88		53.3	11.94		46.8	6.87		40.3	3.98
66.2	44.8		59.7	22.65		53.2	11.82		46.7	6.82		40.3	3.95
66.1	44.2		59.6	22.41		53.1	11.70		46.6	6.76		40.2	3.91
66.0	43.7		59.5	22.20		53.0	11.59		46.5	9.70		40.1	3.88
65.9	43.2		59.4	21.96		52.9	11.48		46.4	6.65		40.0	3.85
65.8	42.8		59.3	21.76		52.8	11.37		46.3	6.59		39.9	3.82
65.7	42.3		59.2	21.54		52.7	11.25		46.2	6.54		39.8	3.79
65.6	41.8		59.1	21.32		52.6	11.13		46.1	6.48		39.7	3.76
65.5	41.3		59.0	21.10		52.5	11.04		46.0	6.42		39.6	3.73
65.4	40.8		58.9	20.86		52.4	10.92		45.9	6.36		39.5	3.70
65.3	40.4		58.8	20.66		52.3	10.83		45.8	6.31		39.4	3.67
65.2	40.0		58.7	20.45		52.2	10.72		45.7	6.26		39.3	3.64
65.1	39.5		58.6	20.22		52.1	10.62		45.6	6.21		39.2	3.61

TABLE 7-10 (cont.)

RCS	L	RCS	L	RCS	L	RCS	L	RCS	L
65.0	39.1	58.5	20.04	52.0	10.52	45.5	6.16	39.1	3.59
64.9	38.7	58.4	19.85	51.9	10.42	45.4	6.10	39.0	3.56
64.8	38.3	58.3	19.66	51.8	10.34	45.3	6.05	38.9	3.53
64.7	37.9	58.2	19.45	51.7	10.22	45.2	6.00	38.8	3.50
64.6	37.4	58.1	19.28	51.6	10.13	45.1	5.95	38.7	3.48
64.5	37.0	58.0	19.09	51.5	10.04	45.0	5.89	38.6	3.45
64.4	36.6	57.9	18.90	51.4	9.96	44.9	5.84	38.5	3.43
64.3	36.2	57.8	18.70	51.3	9.87	44.8	5.80	38.4	3.40
64.3	35.8	57.7	18.52	51.2	9.79	44.7	5.74	38.3	3.37
64.1	35.4	57.6	18.35	51.1	9.70	44.6	5.69	38.2	3.35
64.0	35.0	57.5	18.16	51.0	9.62	44.5	5.65	38.1	3.33
63.9	34.7	57.4	17.98	50.9	9.55	44.4	5.60	38.0	3.30
63.8	34.3	57.3	17.80	50.8	9.48	44.3	5.55	37.9	3.28
63.7	33.9	57.2	17.63	50.7	9.40	44.2	5.50	37.8	3.26
63.6	33.6	57.1	17.45	50.6	9.34	44.1	5.46	37.7	3.23
63.5	33.2	57.0	17.30	50.5	9.26	44.0	5.41	37.6	3.20
63.4	32.9	56.9	17.09	50.4	9.18	43.9	5.36	37.5	3.18
63.3	32.5	56.8	16.94	50.3	9.11	43.8	5.32	37.4	3.16
63.2	32.2	56.7	16.78	50.2	9.04	43.7	5.28	37.3	3.14
63.1	31.9	56.6	16.60	50.1	8.97	43.6	5.23	37.2	3.12
63.0	31.6	56.5	16.43	50.0	8.90	43.5	5.18	37.1	3.10
62.9	31.2	56.4	16.30	49.9	8.84	43.4	5.13	37.0	3.08
62.8	30.9	56.3	16.11	49.8	8.77	43.3	5.09	36.9	3.06
62.7	30.6	56.2	15.96	49.7	8.70	43.2	5.05	36.8	3.04
62.6	30.3	56.1	15.80	49.6	8.63	43.1	5.01	36.7	3.02
62.5	30.0	56.0	15.63	49.5	8.57	43.0	4.96	36.6	3.00
62.3	29.7	55.9	15.49	49.4	8.49	42.9	4.92	36.5	2.98
62.3	29.4	55.8	15.33	49.3	8.43	42.8	4.88	36.4	2.96
62.2	29.1	55.7	15.17	49.2	8.35	42.7	4.84	36.3	2.94
62.1	28.8	55.6	15.03	49.1	8.29			36.2	2.92

Courtesy of the Illuminating Engineering Society of North America.

example problems

Example Problem 1: Point-by-Point

Using the luminaire in Table 7-1 in the data section, calculate the initial illumination at points A, B, C, and D in the figure.

Cross-section   Plan (top view)

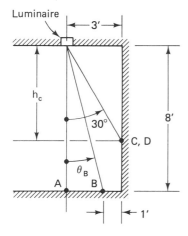 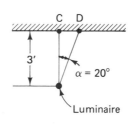

Solution:

Point A:

$$I_0 = 1020 \text{ cd}$$

$$E_A = \frac{1020}{(8)^2} = 15.94 \text{ fc}$$

Point B:

$$\theta_B = \arctan\frac{2}{8} = 14.04° \quad \cos\theta_B = 0.970$$

$$I_{\theta_B} = 6128 \text{ cd}, \qquad D^2 = (8.25)^2 = 68$$

$$E_B = \frac{6128}{68}\cos 14.04° = 87.4 \text{ fc}$$

Point C:

$$h_C = \frac{3}{\tan 30°} = 5.20' \qquad d = \frac{3}{\sin 30°} = 6'$$

$$I_{30} = 10,000 \text{ cd}$$

$$E_C = \frac{10,000}{(6)^2}\sin 30° = 138.9 \text{ fc}$$

195

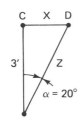

Point D:

$$z = \frac{3}{\cos 20°} = 3.19'$$

$$x = 3 \tan 20° = 1.09'$$

$$d^2 = \overline{3.19}^2 + \overline{5.2}^2 = 37.22'$$

$$\theta_D = \arctan \frac{3.19}{5.2} = 31.53° \qquad \sin \theta_D = 0.523$$

$$\sin \theta_H = \frac{1.09}{6.10} \qquad \theta_H = 10.29° \quad I_\theta = 9240$$

$$E_D = \frac{I_{\theta H}}{D^2} \sin \theta_D \cos \alpha$$

$$= \frac{9240}{37.22} \times 0.523 \times 0.940 = 122.05 \text{ fc}$$

Example Problem 2: Lumen Method—Zonal Cavity

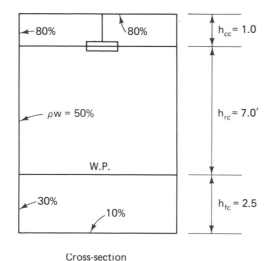

Cross-section

Luminaire Table 7-3

Room dimensions
30 x 35

E = 50 fc

LDD = 0.83 (clean and 24 months)

LDD = 0.87 (6 hrs/ start)

2F40 CW @ 3110 lumens each.

Solution:

$$RCR = \frac{5\,h_{rc}\,(W+L)}{W \times L} = \frac{5(7)\,(30+35)}{30 \times 35} = 2.17$$

$$CCR = \frac{5\,h_{cc}\,(W+L)}{W \times L} = \frac{5(1)\,(30+35)}{30 \times 35} = 0.31$$

Find ρ_{cc} by interpolation for CR = 0.80, CWR = 0.80, and CCR = 0.31 from Table 7-4.

$$\rho_{cc} = 78\%$$

Find CU by interpolation for RCR = 2.17, ρ_{cc} = 0.78, and pw = 0.50, using the luminaire in Table 7-3.

$$CU = 0.62 \quad \text{for} \quad \rho_{fc} = 0.20$$

$$FCR = \frac{5\,h_{fc}\,(W+L)}{W \times L} = \frac{5(2.5)\,(30+35)}{30 \times 35} = 0.77$$

Find actual p_{fc} by interpolation for FR = 0.10, FWR = 0.30, and FCR = 0.77 from Table 7-4.

$$\rho_{fc} = 10\%$$

Find floor correction by interpolation for RCR = 2.17, p_{cc} = 0.78, and pw = 0.50 from Table 7-5.

$$\text{correction factor} = \frac{1}{1.06} = 0.94$$

Corrected CU:

$$CU = 0.62 \times 0.94 = 0.58$$

$$\text{no. of luminaires} = \frac{E \times A}{\text{no. lamps} \times \text{lumens per lamp} \times CU \times LLD \times LDD}$$

$$= \frac{50 \times (30 \times 35)}{2 \times 3110 \times 0.58 \times 0.87 \times 0.83}$$

$$= 20$$

Example Problem 3: ESI Calculation

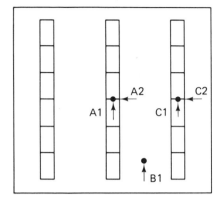

Room characteristics
28' x 28'
F. to clng. = 8' - 0''
WP' = 2.5'
ρ_c = 85%
ρ_w = 48%
ρ_f = 18%

ρ_o = 0.808 for the given task

Given the following preliminary layout investigate the ESI values at three test positions, at 25° viewing angle with two different diffusers.

Data Supplied by a Time-Sharing Terminal

Luminaire types	Position	E_t	L_b	CRF
X	A1	112	95.2	0.804
	A2	141	109.7	0.961
	B1	121	97.4	1.016
	C1	102	87.2	0.778
	C2	137	107.6	0.945
Z	A1	71	56.0	0.942
	A2	70	55.2	0.995
	B1	106	82.7	1.027
	C1	55	44.7	0.899
	C2	64	52.3	0.988

General Equations

$$RCS_e = RCS \times CRF$$

$$RCS_e \rightarrow L_e$$

$$ESI = L_e/\rho_o$$

where RCS_e = effective relative contrast sensitivity, which describes the overall level of visual sensitivity and hence one's ability to perform a visual task

RCS = relative contrast sensitivity, which is a measure of the visual sensitivity of an observer in a luminous environment (Table 7-10)

CRF = contrast rendition factor, which describes the effects of the physical characteristics of the environment on the visibility of a task

L_e = effective luminance, which describes the effects of the environment and state of the visual system on the luminance of the task

198

ESI = equivalent sphere illumination, which is a measure of the level of illumination under reference conditions that would produce task visibility that is equivalent to the visibility produced in a real environment

ρ_o = task reflectance measured in a photometric sphere

E_t = task illumination, which is a measure of the total illumination on the surface of the task

L_b = task background luminance, which is a measure of the background luminance of the task

Calculation of ESI

Luminaire Type ___X___   Position___A1___

1. L_b = ___95.2___ .
2. Find RCS based on L_b from Table 7-10.

$$RCS = \underline{\quad 73.1 \quad}$$

3. CRF = ___0.804___ .
4. Compute RCS_e = RCS × CRF.

$$RCS_e = \underline{\quad 73.1 \quad} \times \underline{\quad 0.804 \quad} = \underline{\quad 58.8 \quad}$$

5. Find L_e based on RCS_e from Table 7-10.

$$L_e = \underline{\quad 20.6 \quad}$$

6. ρ_o = ___0.808___ .
7. Compute ESI = L_e/ρ_o.

$$ESI = \underline{\quad 20.6 \quad} \div \underline{\quad 0.808 \quad} = \underline{\quad 25.5 \quad}$$

8. Enter ESI value in the worksheet.

Luminaire Type ___X___   Position___A2___

1. L_b = ___109.7___ .
2. Find RCS based on L_b from Table 7-10.

$$RCS = \underline{\quad 74.3 \quad}$$

3. CRF = ___0.961___ .
4. Compute RCS_e = RCS × CRF.

$$RCS_e = \underline{\quad 74.3 \quad} \times \underline{\quad 0.961 \quad} = \underline{\quad 71.4 \quad}$$

5. Find L_e based on RCS_e from Table 7-10.

$$L_e = \underline{\quad 78.9 \quad}$$

6. ρ_o = ___0.808___ .

7. Compute ESI $= L_e/p_o$.

$$\text{ESI} = \underline{\quad 78.9 \quad} \div \underline{\quad 0.808 \quad} = \underline{\quad 97.6 \quad}$$

8. Enter ESI value in the worksheet.

Worksheet

Lens system	Calculation headings	Position				
		A1	A2	B1	C1	C2
X	ESI fc	25.5	97.6	137.5	19.9	85.3
	E_t, measured fc	112.0	141.0	121.0	102.0	137.0
	LEF $=$ ESI/E_t	0.23	0.69	1.14	0.20	0.62
Z	ESI fc					
	E_t, measured fc LEF $=$ ESI/E_t	71.0	70.0	106.0	55.0	64.0

8 lighting and energy conservation

Lighting has been one of the prime targets of mandatory standards to reduce energy consumption. To put the role of lighting as it relates to energy conservation into perspective, we shall examine the impact of lighting on total energy resource consumption in the United States. The goal of energy conservation should be to reduce the consumption of energy resources.

Currently, 80 percent of the resources used in this country are fossil fuels (coal, oil, and natural gas)[1]. The most critical fuels in terms of estimated reserves are oil and natural gas. Of the total resources consumed, approximately 25 percent are used to generate electricity. Twenty percent of that 25 percent ends up as lighting. Thus only approximately 5 percent of the total energy resources consumed in this country end up in the form of lighting. Approximately 9 percent of the 25 percent used to generate electricity involves oil and natural gas. That is, only about 3 percent of the total energy resources used to generate electricity involve critical fuels.

With these facts, the question must be asked, why is lighting a target for energy conservation? Lighting is "visible." Second, in terms of the end user, lighting represents 30 to 50 percent of the operating cost of a building. Lighting energy conservation is important in terms of the total resource reserves and in terms of operating cost for the building owner. As utility rates continue to increase, the impact of lighting on operating cost will become painfully apparent. Make no mistake, much lighting is wasteful. Qualified lighting engineers have known for years that high levels of uniform lighting throughout a space are wasteful. For example, the *IES Lighting Handbook*[2] recommends 100 fc (Figure 9-80, p. 9-81) at the *task* for general office work involving "hard pencil or poor paper reading fair reproductions." The problem facing the engineer is task definition and location. Since neither of these design parameters is known during a typical design process[3], the designer blankets the entire space with 100 fc. This uniform approach is not recommended by the IES, but has resulted from a lack of communication and understanding of fundamentals on the part of the engineer and architect. The *IES Lighting Handbook*[2] (footnote, Figure 9-80, p. 9-81) states that "illumination levels shown in the table are intended to be minimum on the task."

The oil embargo of 1973 brought the "energy crisis" to a head. One of the first documents that was a direct outcome of the energy crisis and the visibility of lighting was a General Services Administration/Public Building Services (GSA/PBS) document. The GSA/PBS guidelines entitled "Energy Conservation Guidelines for Existing Office Buildings" was completed in early 1974. At approximately the same time a document entitled "Design and Evaluation Criteria for Energy Conservation in New Buildings" was prepared by the National Bureau of Standards (NBS) for the National Conference of States on Building Codes and Standards (NCSBCS) on lighting for new construction. NCSBCS asked the American Society of Heating, Refrigerating, and Air Conditioning Engineers (ASHRAE) to prepare a standard for energy conservation in new buildings based on the NBS document.

In the spring of 1974, the Illuminating Engineering Society (IES) was asked by ASHRAE to prepare Chapter 9 on lighting. A revised format of Chapter 9 was prepared and sent to ASHRAE in May 1974. The May 1974 revised version was different from the earlier version in that it established a procedure for determining a **lighting power budget.** After numerous revisions, public review, and several meetings of the IES Task Committee on Energy Budgeting Procedures, a Lighting Power Budget Determination Procedure (Chapter 9) was approved by the board of directors of the IES in July 1975. In late 1975, ANSI/IES ASHRAE Standard-90P (preliminary) was approved and became ANSI/IES ASHRAE Standard 90-75.

In 1975, Congress passed an act entitled "Energy Policy and Conservation Act of 1975," public law 94-163, which was amended by public law 94-385 ("Energy Conservation Standards for New Buildings Act of 1976"). Public law 94-163 contains mandatory lighting efficiency standards for public buildings; they must meet the standards outlined in Tables 8-1 and 8-2.

TABLE 8-1

Public Law 94-163

Mandatory lighting efficiency standards:

1. Be in place by 1/1/78.
2. Apply to *all* public buildings.
3. Be no less stringent than Section 9 of Standard 90-75.
4. For existing buildings contain elements deemed appropriate by the state.

TABLE 8-2

Public Law 94-385

Energy conservation for new building requires:

1. Federal action to ensure energy conservation features in new buildings.
2. Development and implementation by 1/1/80 (performance standards).
3. Performance standards to maximum practicable energy efficiency.
4. Encourage states and local governments to adopt and enforce the standards.

In 1976, the Energy Research and Development Administration (ERDA) contracted with NCSBCS to codify ASHRAE 90-75. The resulting document was called "The Model Code for Energy Conservation in New Buildings." Section 9 of the model code is based on section 9 of ASHRAE 90-75. The NCSBCS model code has been adopted by a number of states to satisfy the requirements of public law 94-163 and 94-385.

In June 1976, the board of directors of the IES adopted important revisions to Chapter 9, which are incorporated in ASHRAE 90-75R (revised). These revisions will tighten the power budget procedure and assure energy conservation. The NCSBCS model code does not reflect these important revisions. The illuminating engineer needs to keep current with the latest revisions and make certain that the correct document is being used.

ASHRAE 90-75R was cosponsored by IES and ASHRAE and submitted to the American National Standards Institute (ANSI) in late 1977 for adoption as an ANSI standard. The resulting document is known as ANSI/IES/ASHRAE standard 90, "Energy Conservation in New Buildings."

In 1975, work was begun on a series of six documents (100 series) that deals with "Energy Standards for Existing Buildings." The documents cover low-rise residential (ASHRAE 100.1/IES-EMS 4.1), high-rise residential (ASHRAE 100.2/IES-EMS 4.2), commercial (ASHRAE 100.3/IES-EMS 4.3), industrial (ASHRAE 100.4/IES-EMS 4.4), institutional (ASHRAE 100.5/IES-EMS 4.5), and public assembly occupancies (ASHRAE 100.6/IES-EMS 4.5). The documents are going through a concentrated review process that should result in a more rapid adoption as ANSI standards.

ILLUMINATING ENGINEERING SOCIETY ENERGY CONSERVATION RECOMMENDATIONS

In 1972, the IES prepared 12 recommendations[4,5] for better utilization of energy in lighting design without sacrificing quality. The recommendations cover the operation, maintenance, and selection of lighting equipment. The recommendations apply to new construction as well as renovations or retrofit (upgrade). The 12 recommendations are as follows:

1. Design lighting for expected activity (light for seeing tasks, with less light in surrounding nonworking areas).
2. Design with more effective luminaires and fenestration (use systems analysis based on life cycle).
3. Use efficient light sources (higher lumen per watt output).
4. Use more efficient luminaires.
5. Use thermally controlled luminaires.
6. Use lighter finish on ceilings, walls, floor, and furnishings.
7. Use efficient incandescent lamps.
8. Turn off lights when not needed.
9. Control window brightness.
10. Utilize daylighting as practicable.
11. Keep lighting equipment clean and in good working condition.
12. Post instructions covering operation and maintenance.

To expand on these recommendations,

1. Expected activity: a design approach that uses lower ambient levels with higher task levels will produce an energy-efficient system. The task location must be known in order to supply the appropriate lighting level at the task location.
2. More effective luminaires and fenestration: luminaires and fenestration

systems should be as efficient as possible without adversely affecting comfort (VCP) and visibility (ESI).

3. Efficient light sources: selection should be based not only on the efficacy (lumens/watt), but also on the life, cost, and color rendition. Color should carry as much weight as the other factors, since color has a direct bearing on the psychological behavior of people, which affects productivity and mood.

4. More efficient luminaires: efficiency includes the utilization of energy in the space as well as cleaning and relamping convenience.

5. Thermal control: make use of the heat produced by the lighting equipment.

6. Reflecting surfaces: the absorption of light in a space due to low reflectance values will reduce the efficiency of the lighting system, resulting in an increase in wattage.

7. Efficient incandescent lamps: if incandescent lamps must be used, the higher-wattage lamps will be slightly more efficient than lower-wattage lamps.

8. Lights off when not needed: switching off lights when not in use will result in energy savings and a reduction in operating costs. The effectiveness of the energy reduction is a function of the flexibility of the control systems.

9. Window brightness: excessive glare of windows on the line of sight will affect comfort and visibility, which may reduce performance.

10. Daylighting: the effectiveness of daylight is dependent on the combined daylight/artificial lighting system. Unless the control system is properly designed, energy reduction is questionable.

11. Maintenance: good maintenance will require fewer luminaires by increasing the utilization of the light entering the space. Maintenance should include not only spot relamping, but also room surface cleaning, painting, luminaire cleaning, and group relamping.

12. Operating and maintenance instructions: the design of sophisticated lighting systems and controls for energy conservation will be wasteful if the building user does not know how to properly use and maintain the system. The designer should provide instructions on how to use the system.

POWER, POWER FACTOR, AND ENERGY

The goal of good lighting design is to save energy and, at the same time, to maintain visual performance and productivity. The current emphasis in energy conservation is on the reduction of power (fewer lamps) by reducing lighting levels. This approach leads to design techniques based on watts per square foot or power design techniques.

Power

Power (true power) is measured with a wattmeter and defined as

$$P = E \times I \times \cos \theta$$

where E = voltage in volts
   I = current in amperes
   $\cos \theta$ = power factor

Power Factor

If the power factor is unity or 100 percent, the voltge vector will coincide with the current vector, and the resulting power will be the product of the voltage and current:

$$P = E \times I \quad \text{volt-amperes}$$

The voltage and current are said to be in phase, which means that the cosine of the angle θ is 1.0. Incandescent lamps are pure resistance loads with a unity power factor.

All gaseous discharge sources require a ballast that contains windings that produce an inductive reactance. Inductive reactance causes a phase shift between the voltage and current, resulting in a phase angle θ greater than 0 or a power factor of less than 1. Inductive reactance causes the current to lag behind the voltage or creates a lagging power factor, which results in wasted or nonproductive power. When the power factor is less than unity, nonproductive current is present, which (1) requires larger conductors (increased wire cost), (2) adds to the voltage drop (increased copper losses), (3) reduces the circuit's ability to deliver power (increased power or energy consumption), (4) overloads transformers, panels, feeders, and cables, and (5) results in power factor penalty billing. The power factor is numerically equal to the cosine of the angle θ (see Figure 8-1).

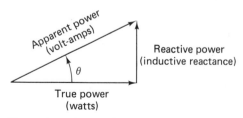

Reactive power
(inductive reactance)

True power
(watts)

Figure 8-1.   Power Triangle

$$\text{power factor} = \frac{\text{true power}}{\text{apparent power}}$$

$$= \frac{\text{volts} \times \text{amperes} \times \text{cosine } \theta}{\text{volts} \times \text{amperes}} = \text{cosine } \theta$$

$$= \frac{\text{wattmeter}}{\text{voltmeter} \times \text{ampmeter}}$$

206

The apparent power is the product of the voltage and current and is expressed in volt-amperes. The typical utility meter is a watt-hour meter that measures the true power. The generating capacity must be sufficient to deliver the apparent power, which can result in a sizable waste of power (energy) if the angle θ is large. To reduce this wasted power (energy), the power factor should be corrected to 0.90 (90 percent). Lagging power factor is usually corrected by placing a capacitor in the circuit to reduce the phase angle.

Energy

The goal is energy conservation. Energy is the product of power (true power) used by a load and the amount of time that the load is in use. Energy is measured with a watt-hour meter. In other words, to be energy minded, not only the amount of electricity used (power) but also how long it is used (time) must be taken into account:

$$\text{energy} = \text{power} \times \text{time}$$

To reduce energy consumption by the lighting system, the illuminating engineer must go beyond simple lighting layouts (lighting design) and study the control of the lighting system.

Spielvogel[6] points out a number of facts collected by various agencies which led him to state that "the one factor that, more than any other, determines energy consumption of a building is how it is used." In particular, a plot of the installed lighting watts per square foot versus the total lighting energy consumption of buildings across the country shows no correlation (Figure 8-2). If there is no correlation between designed watts per square foot and total energy used, what are the factors involved in total energy consumption? Most experts agree that one factor is the control of the system.

LIGHTING CONTROLS

The numerous methods to control lighting energy consumption fall into two basic categories. The first type of control provides for either an **on** or **off** state; the second category provides on–off control, but in addition provides the ability to select a level of energy consumption between on and fully off. This chapter discusses on–off controls, level controls, and the state of the art of a combination of the two.

On-Off Controls

The basic on–off control is the switch. Switches are available in a number of configurations, each suited to a particular function.

Circuit-Breaker Switching

The National Electric Code requires all branch circuits to have circuit protection. Most branch circuit protection today is accomplished with circuit

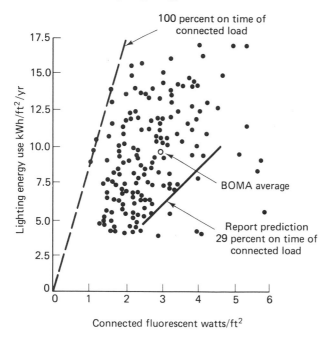

Figure 8-2. Office Building Lighting Energy Consumption versus Installed Watts. (Courtesy of L. Speilvogel)

breakers. To save initial costs, many designs use the circuit breaker to switch the load. Although the initial cost of a switch leg is saved, circuit breakers should not be used in place of switches. State energy codes that have adopted the NCSBCS-model energy codes require a minimum of one switch per circuit, which makes this practice illegal in some states. In addition, the use of circuit breakers as switches is energy inefficient. Most often the location of the panel board is inaccessible to the occupants of the space being served. In such cases, the occupants have no control over their lights, and as a consequence the lights will burn even when not needed. Also, a 20-A branch circuit loaded to maximum capacity according to code can handle 9 four-lamp fluorescent luminaires at 120 V or 22 four-lamp fluorescent luminaires at 277 V. In large areas, all the units may be needed at one time, but in offices or small areas this approach results in more than one space being served by a single circuit breaker. This will prevent energy reductions, because it keeps individual office occupants from turning off the lights as they leave the space. Although this method is often used, it is not recommended and should be avoided if energy savings are to be realized.

AC Snap Switches

The most common on–off device is the ac snap switch. The ac snap switch can be located almost anywhere within the room and can carry the full branch circuit load. For example, each private office should have at least one switch for the lights within that space, and larger areas can be broken up into distinct work areas with a switch in each area. For a space with more than one entrance, three-way or four-way switches should be used to provide control at each entrance.

When a number of switches are provided to control the lights in a large area, they should be circuited with the function of the space dictating the configuration. As an example, consider a lecture hall in which projection equipment is often used. One pattern of circuiting is shown in Figure 8-3(a). With switching circuited in this way, there are always lights on in the front of the room. Figure 8-3(b) is a more logical switching pattern that allows the lights in the rear of the room to remain on for taking notes while those in the front can be turned off to increase the contrast of the screen image. Open plan or office-scape designs are

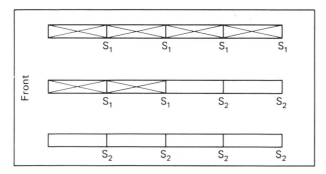

(a) Switching diagram A

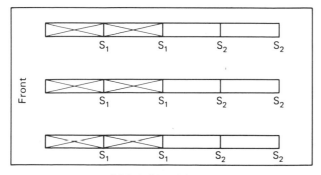

(b) Switching diagram B

Figure 8-3. Switching Based on Room Function and Utilization

209

becoming more popular because of the increased space utilization and flexibility. Separate controls can be provided for each indirect luminaire and task light at each luminaire location. This allows for selective control of ambient lighting as well as task lighting as the space utilization changes during a day.

Although the ac snap switch is inexpensive and the most commonly installed control device, it has many disadvantages because it carries line voltage (120 or 277 V). Although its safety record is good, the possibility of death or serious injury is greater at these voltages. Also, to place the switch in its optimum functional location can be economically prohibitive because of wiring costs, as well as increased voltage drop due to long runs. The length of runs and cost become more of a factor with the use of three- or four-way switches.

As an example of both dollar savings and energy savings, consider four offices with 2 four-lamp luminaires per office, each office used 9 h/day, 5 days/week. In one case, there is one switch controlling all the offices; in the other, there is one switch per office. The second alternative makes the assumption that each person is out of his office 3 h/day and that he will turn off the lights whenever he leaves. This second assumption is optimistic, since most studies indicate that people do not turn off lights when leaving a room.

	Case 1	Case 2
Initial Cost	One switch and wiring: $13	Four switches and wiring: $52
Annual electricity consumed	× 8 luminaires	× 8 luminaires
	× 200 W/luminaire	× 200 W/luminaire
	× 9 h/day	× 6 h/day
	× 260 days/year	× 260 days/year
	3744 kWh/year	2496 kWh/year

It costs $39 more for four switches and associated wiring, but saves 1248 kWh/year in electrical usage. The cost of electrical power would have to be 3.125¢/kWh to recover the cost of the added switches in 1 year. In many areas of the country, the cost of electricity is far above this rate; therefore, case 2 would be more economical than case 1.

Time Switches

In the preceding example the assumption was made that a person would turn off the lights in his space whenever he left the room. This is not always true, so it has been suggested that systems be used that take over the turning off of lights. One such device is the time switch, most often used with sunlamps installed in bathrooms. One manufacturer provides models with timed cycles of 0 to 5, 0 to 15, 0 to 30, and 0 to 60 min, and 0 to 6 and 0 to 12 h. They also have 0 to 3 min, 0 to 60 min, and 0 to 12 h times that feature a "hold" position for leaving the circuit closed continuously. The price is reasonable at about $10 each.

With a time switch, the occupant of a space must actively turn on the lights but has to take no action to turn them off. Since the behavior of individuals related to the control of lighting is not really known, the economics of the time switch should be compared to both the single switch control of four offices and individual control in each office. In the previous analysis, it was assumed that 3h/day would be spent out of the office. It is probable that part of that time is for short durations in which a time switch would remain on. Therefore, the time out of the office will be reduced to 2 h with the use of time switches.

	Case 3	
Initial cost		One switch and wiring: $13
Annual		1600 W
electricity	×	9 h/day
costs	×	260 days/year
	×	4¢/kWh (assumed) = $149.76

	Case 4	
Initial cost		Four switches and wiring: $52
Annual		1600 W
electricity	×	9 h/day
costs	×	260 days/year
	×	4¢/kWh (assumed) = $149.76

	Case 5	
Initial cost		Four time switches and wiring: $92
Annual		1600 W
electricity	×	7 h/day
costs	×	260 days/year
	×	4¢/kWh (assumed) = $116.48

Using a 25-year economic life and 10 percent cost of money, the saving investment ratio is (SIR) calculated:

$$SIR = \frac{\text{present value of future savings}}{\text{present value of incremental investment cost}}$$

$$= \frac{\text{present value factor} \times \text{yearly savings}}{\text{present value of incremental investment cost}}$$

The SIR of case 5 to case 3 is

$$SIR_{31} = \frac{9.524 \times \$33.28}{\$79} = 4.0$$

and the SIR of case 5 to case 4 is

$$SIR_{32} = \frac{9.524 \times \$33.28}{40} = 7.9$$

Any SIR greater than 1 economically justifies the added investment. In this situation, it is economical as well as energy efficient to install time switches when the cost of electricity is 4¢/kWh. There are other spaces, such as warehouses, bathrooms, janitor closets, or storerooms where personnel enter the space for short periods of time and neglect to turn off the lights. Time switches would prove even more advantageous in these applications. Where conventional (snap) switches are installed and used, the additional cost of installing time switches is obviously not warranted.

The major disadvantage of time switches is the audible ticking of the timing mechanism. In small areas, such as offices, this could be annoying. It would be counterproductive to improve the luminous environment and adversely affect the acoustical environment.

Low-Voltage Switching

Low-voltage remote switching is not new on the market, but is gaining in importance. All low-voltage systems consist of a magnetic relay, a transformer (putting out 24 V or less), and switches that are interconnected with low-voltage wiring. The relay switches the line voltage and current with a low-voltage command from a low-voltage switch in the space. This provides the ability to control loads from great distances, control a number of different loads from one location, and control one load from multiple locations. Two pamphlets printed by manufacturers of low-voltage equipment, "Low Voltage Remote Control Switching" from General Electric, Wiring Device Business Department, Providence, Rhode Island 02940, and "Low Voltage Lighting Control" from Robertshaw Controls, 1800 Glenside Drive, Richmond, Virginia 23226, are recommended for specific circuit suggestions.

A low-voltage system has many advantages. Since all switch legs carry 24 V or less, there is greater safety, lower copper costs (smaller guage wires), and lower installation costs because conduit or a raceway system is not required in most states. Switches for a space can be placed in many locations, such as the supervisor's office or guard station. Flexibility can be achieved easily with low-voltage control. Each luminaire or any group of luminaires up to branch circuit capacity can be grouped on a relay. The switching pattern then can be put together by connecting the low-voltage wiring and switches as needed.

One possible configuration would be to switch off all the office lights in a building just after work hours. Then if anyone desires to stay and work overtime, they may turn on the lights again using a low-voltage switch located in the area. With the advent of computers to control building functions, the use of low-voltage control of lighting will become more important. One manufacturer is developing computer interface modules.

Low-voltage remote-control switching becomes more attractive economically as the size and complexity of the control system increase. The initial cost of a low-voltage system may be higher than conventional methods. The initial cost is dependent on the scope and complexity of the design. The transformer, relay, wire, and switch costs may be offset by reduction in wire and installation costs. The selection of a low-voltage system based on a lower initial cost does not take into account the savings in electricity because of added control.

Time Clocks and Photocells

The controls previously discussed in this chapter require human action to initiate a change of state. Time clocks and photocells, on the other hand, do not. Time clocks and photocells are discussed together because they are conventionally used to control outdoor lighting, such as security, parking lot, roadway, and area lighting. Other than initial cost, there is no reason for not using these systems inside. There are a number of time clocks and photocells that provide different degrees of control.

The most elementary time clock has a single on time and a single off time each day of the week. It has no provision to adjust for seasonal changes in sunrise and sunset, and thus must be continually adjusted when used in outdoor applications. For indoor applications, such as offices, there is usually no need to adjust on–off times, but on weekends it is desirable not to turn on the lights. To overcome the disadvantage of continuous operation over the weekend, a time clock that is able to skip a day is available. One model provides for skipping days, but still has a single on and a single off time for the days not skipped.

If a different on or off time is needed for certain day and/or days need to be skipped, a 7-day time clock should be used. Again, seasonal changes in dawn and dusk must be manually compensated for when used in outdoor applications. Designed to take into account seasonal changes in the day–night cycle, astronomic clocks that have a special driving gear are available. No adjustments are needed, but they must be ordered for specific locations. Besides turning off at dawn, the clock can be set to turn off anywhere from 3:30 P.M. to 2:30 A.M. An example of this feature would be a parking lot that is never used after a certain hour at night.

Rather than using time to determine when lights should be on or off, photocells or photocontrols can be used to sense when the light level decreases (sunset) or increases (sunrise). Some photocontrols are constructed so that the level at which they are activated can be varied. This is done simply by a movable plate across the cell face. Photocells can be used alone or in conjunction with other devices. To provide lighting at sunset that is not needed all night, a photocell can be used to turn on the lights with a time control that would turn them off at a preset time. When interfaced with low-voltage remote switching, photocells provide another means of controlling the lighting system and its energy consumption. Photocells are an excellent choice of control because they are simple, maintenance free, independent of a source of operating power, and self-adjusting to seasonal changes.

Ultrasonic Detectors

Whether control of the lights should be fully in the hands of the people using the space is debatable. If everyone were energy minded and always remembered, there would be no need for devices that work independently of human interaction. But this generally is not the case. A device that operates independently of human action, although not used in lighting presently, is the ultrasonic detector. Designed and used for security (alarm) systems, the device uses ultrasonic waves to detect the presence (by movement) of persons within the room. Obviously, the development of such a device to switch lighting would completely remove the question of human behavior and result in optimum energy utilization. A person entering a room would cause the lights to come on and remain on until the person leaves. This is control of the lighting by the mere presence of an occupant. A switch that would override the ultrasonic and provide for off control would be needed only in rooms where projection equipment might be used. The units that are on the market now are used for intruder detection and are expensive. With the increased cost of energy and decreasing availability, an ultrasonic switch could prove economically justifiable.

Level Controls

The first part of this chapter discussed a variety of means and devices for turning lights on and off. Many circumstances exist where a level of artificial illumination somewhere between fully on and off is desired.

Dimmers

The best means of controlling the level of illumination is the dimmer. The original dimmers were resistance types that diverted some of the current through a variable resistor. Although the resistance dimmers have their advantages and disadvantages, the only characteristic of importance here is that they do not save energy. The total power used is the same whatever the light level, because the dimmer itself draws power. Most existing dimmers of this type are used exclusively in theatrical applications.

In the last 10 years, the solid-state dimmer has taken over 90 percent of the dimmer market. The principle of operation is simple. The current in an ac power system is a periodic function, as depicted in Figure 8-4. The power delivered to the load is a function of the shaded area under the curve. The electronic dimmer, by means of an electronic switch, turns off the current to the load for a portion of the cycle, as in Figure 8-4. Note that the shaded area under the curve is smaller, and thus less power is being delivered to the load. The electronic switch has a continuously variable operating time from fully on to fully off. Since the current cannot come on instantaneously, as the vertical line would suggest in Figure 8-4, there is a slight slope or rise time. The current will also overshoot the ac current curve a bit, after which it will oscillate a short time. The overshoot and oscillation cause radio-frequency (RF) noise and lamp filament ringing. Most manufacturers provide for RF suppression and have chokes available that reduce filament ring.

Solid-state dimmers are now made for incandescent, fluorescent, and HID lighting. Incandescent dimmers, which only require the electronic switch and RF

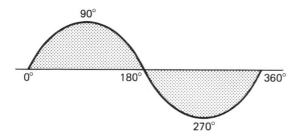

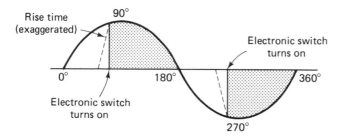

Figure 8-4. Dimmer Characteristics for Alternating Current

suppression, are available in a wide range of power capacities. Wall box units are available from 600 to 2000 W. Other systems that use dimmer packages remotely located from the intensity control switch are available in sizes from 100 to 300,000 W. Remote dimmer pack systems are available with multiple remote-control stations or with a control system that allows push-button selection of a number of preset levels, full range control, and override on–off. Incandescent dimming is the only type that can be dimmed from 100 percent of full light output to 0 percent.

Fluorescent dimmers were first introduced on the market about 1962. They are not interchangeable with incandescent dimmers. They require different cir-cuitry, a special dimming ballast, and special lamp holders. Unlike incandescent lamps, when dimmed, fluorescent lamps never go off completely. The minimum light output of any system on the market is $1/500$ of the maximum. This is suffi-ciently low that they appear to be off. Up to 30- 40-W lamps may be dimmed by a single wall-box-mounted dimmer; up to 80 lamps may be controlled by a single dimming module. The dimming module system requires intensity controls to be mounted remotely from the dimming module. Because of this, dimming modules may be ganged or combined with incandescent modules. Remote-control, high-capacity systems may also be outfitted with preset-type controls. A new system introduced by one manufacturer uses an intensity control, dimming module, and electronic package that replaces the ballast. Figure 8-5 shows the increase in ef-ficacy of this new system over the conventional dimming arrangement.

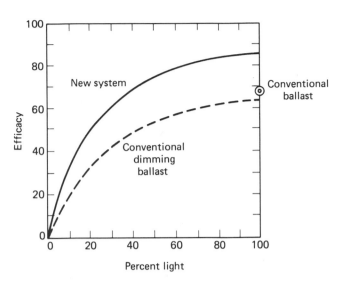

Figure 8-5. Efficacy Versus Light Output of Conventional and New Fluorescent Dimming Systems

With the increased use of HID lamps, it is desirable to have dimming for them. Mercury vapor dimming systems are currently available from three manufacturers. One requires no ballast and as a consequence is much quieter than ordinary systems. Because the arc must be maintained, minimum light output of any system is 2 percent of maximum. The mercury dimming system requires a dimmer module with remote intensity control and is available with a preset control package. Systems for dimming metal halide and high-pressure sodium systems are available.

The dimmer modules used in the high-power fluorescent, incandescent, and HID systems are ideal for interfacing with time clocks, photocells, or computers. Such a system has been developed by one manufacturer for the control of a mercury vapor system. It consists of a photocell, dimmers, and a control station to monitor the illumination level and the dimmers. The selected level of illumination is maintained by adjustments of the dimmers by the controller as changes in lumen output, daylight contribution, and dirt accumulation occur. Development of a similar system for fluorescent or incandescent lighting would be valuable.

Dimming systems for fluorescent and HID lamps are complex and expensive. The problem lies in the inductive ballasts. Electronic, high-frequency ballasts that would make dimming easier and less costly have been developed but not marketed for fluorescent lamps. A simple variable resistor or capacitor added to the circuit is said to allow for simple, easy dimming.

Dimmable lighting systems are expensive, and may not be applicable for every installation; but they will result in energy savings if used properly. Dimmers should be used only where it is anticipated that level control will be needed and used.

Multilevel Ballasts

A second method of level control introduced in the past few years is the multilevel ballast. The purchase of a small number of different items results in an increase in the initial cost of a building. With multilevel ballasts, only one type of luminaire and ballast need be specified for the job. Adjustments in levels of illumination can be made by a simple wiring change. When retrofitting of existing installations is necessary, multilevel ballasts can save money and energy in areas that were overlighted without greatly sacrificing uniformity[7].

Multilevel ballasts are available in two- and three-level models. The two-level ballast provides for 100 percent light output at 98.7-W input using two F40T12 RS lamps and 55 percent light output at 55.7 W. The three-level ballast has 100, 55, and 37.8 percent light output at 98.7, 55.7, and 37 W, respectively. A disadvantage of the multilevel ballast is that it requires an electrician to make the level change by rewiring the ballast. Local codes may not allow a switch to be placed in the level leads of a multilevel ballast. A two- or three-position switch mounted in the luminaire would make illumination level control easier and eliminate the need for an electrician to make the modification.

State-of-the-Art

A paper presented at the 1976 IES Annual Convention[8] describes a computer-controlled lighting control system. It combines remote on–off control with two-level illumination selection. The system consists of a controller and receiver/switch. The controller is a microcomputer and oscillator. The receiver/switch is a decoder and two relay switches. The microcomputer is programmed for different lighting patterns and the addresses of the luminaire. The address and condition codes generated are modulated by the oscillator and superimposed on the building electrical system line frequency. The message travels through the building electric distribution system. At the receiver/switch, a decoder takes the message off the line, and if the address code corresponds to the one given that switch, the condition codes are executed, turning the luminaire fully on, halfway on, or off. A clock in the computer times the events, allowing for different lighting patterns to be executed at different times of the day.

The advantages of a system of this type are manyfold. Because the commands are sent from the controller to the receiver/switch through the power lines, no rewiring is needed when retrofitting, no switch legs are needed, and no control wires in addition to the regular wiring would be needed. If the cost of the receiver/switch is low, each luminaire in a building could be equipped with one. Since each would have its own address, any lighting configuration could be programmed. As the system exists now, only preprogrammed configurations can be executed. With the microcomputer, various inputs can be used, such as time clocks, photocells, and/or remote input devices.

This use of the microcomputer wastes the microprocessor's capability, because it functions only when a change of state is called for. Future developments could include feedback from various other building functions.

Decisions regarding building performance could be made and appropriate adjustments executed. One such function could be the monitoring of peak electrical demand. If demand were exceeding a predetermined level, all building functions would be checked and those not necessary would be shed. The continuing development of microprocessor-controlled lighting will bring about the desired flexibility while cutting costs and saving energy.

LIGHTING POWER BUDGET

A lighting power budget procedure has been established by the Illuminating Engineering Society. The procedure is based on the state of the art in the field of illuminating engineering. It is recognized that the procedure deals with power rather than energy. The illuminating engineer should strive for maximum visual performance and a pleasing environment without exceeding the budget limit. Energy-efficient lighting systems can be achieved when proper lighting controls and effective maintenance procedures are combined with an adequate power budget.

ANSI/IES/ASHRAE Standard 90

Standard 90, "Energy Conservation in New Buildings," establishes a maximum or upper limit on the power to be consumed by the lighting system. It is not a lighting design procedure; it is an analysis procedure. The procedure places only one constraint on the design of the lighting, which is the maximum (total) watts available for lighting. The lighting power budget procedure (see Figure 8-6) is based on task lighting and the use of efficient light sources and equipment. EMS-1R, "IES Recommended Lighting Power Budget Determination Procedure"[9] (Section 9, ANSI/IES/ASHRAE 90), and EMS-3R, "Example of the Use of the IES Recommended Lighting Power Budget Determination Procedure"[9], are available from the Illuminating Engineering Society to aid the engineer in calculating a lighting power budget. Figure 8-6 is provided to show the method of calculation. An example problem can be found at the end of this chapter.

ASHRAE 100 Series/IES-EMS 4 Series

The six ASHRAE 100/IES-EMS 4 standards are used to establish lighting requirements for the conservation of energy in existing buildings. To achieve the objective of lighting-energy conservation, the connected lighting load is determined for the existing building. A lighting power budget is computed for the existing building following the procedure outlined in EMS-4[9]. The connected lighting load is compared against the calculated lighting power budget. If the connected load exceeds the budget figure, a program for effective lighting-energy management must be developed to bring the connected load into compliance with the computed lighting power budget.

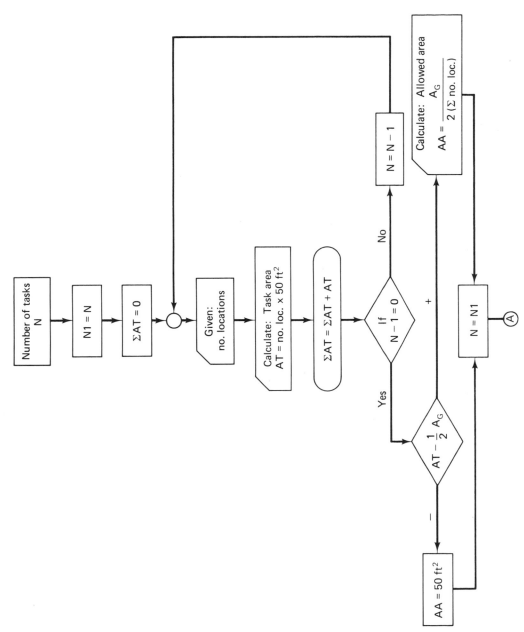

Figure 8-6. Power Budget Procedure Flow Diagram

219

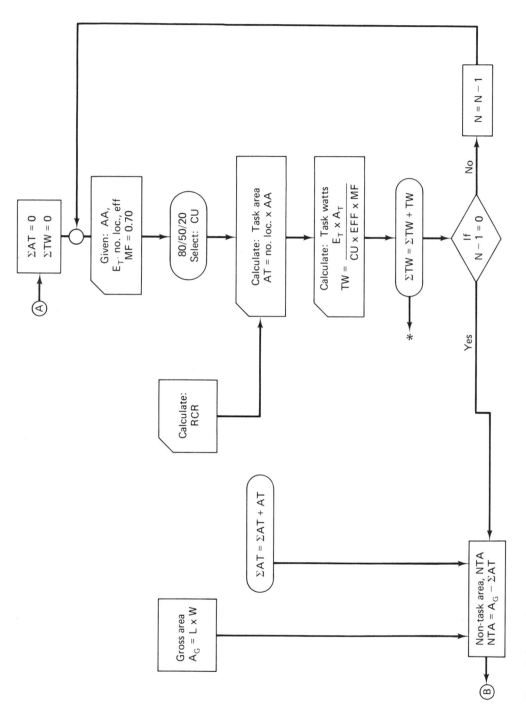

Figure 8-6 (cont.)

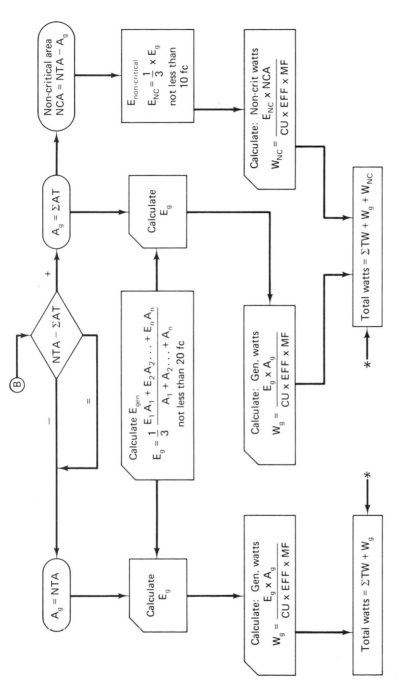

Figure 8-6 (cont.)

221

references

1. Helms, R. N., "Lighting and Energy Conservation—What is Reasonable?," **Electrical Consultant,** May 1975.

2. Kaufman, John, editor, **IES Lighting Handbook,** 5th ed. Illuminating Engineering Society, New York, 1972.

3. Helms, R. N., "Lighting Design Without Analysis: The Consulting Engineer's Dilemma," Proceedings—The Basis for Effective Management of Lighting Energy Symposium, Federal Energy Administration, Oct. 29 and 30, 1975, Washington, D.C.

4. Ringgold, P., "In the Interest of Illumination," **Lighting Design and Application,** Nov. 1972, pp. 1a–6a.

5. Kaufman, J., "Optimizing the Uses of Energy for Lighting," **Lighting Design and Application,** Oct. 1973.

6. Spielvogel, L. G., "Exploding Some Myths About Building Energy Use," **Architectural Record,** Feb. 1976, pp. 125–128.

7. **New Advance Multi-level Ballasts,** produce specification sheet, Advance Transformer Co., Chicago, IL, 1975.

8. McGowan, T. K., and G. E. Feiker, "A New Approach to Lighting System Control," **Journal of the Illuminating Engineering Society,** Oct. 1976.

9. Documents available from Illuminating Engineering Society, 345 East 47th St., New York, N.Y. 10017.

data section

Photometric data

Sketch:

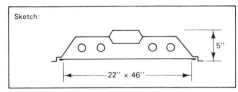

|—— 22" x 46" ——| 5"

Description:

_____ 2 x 4 troffer w/6250 _____

Test No. __1857—1__ Cat. No. __682—4__
Lamps __4F40CW__ Volts __120__
Lumens/lamp __3100__ Test dist. __20__
Shielding angle : N __90__ L __90__
Bare lamp __2350__ Date _____
By __T.B.__ Chkd. by _____

LUMNO = DB117

Luminance data (average)			
Angle	Maximum	Average	Ratio max./ave.
45°	2280	1478	1.5
50°	1900	1145	1.7
55°	1130	772	1.5
60°	1020	542	1.9
65°	890	408	2.2
70°	880	426	2.1
75°	860	440	2.0
80°	840	457	1.8
85°	750	503	1.5
90°	0	0	0

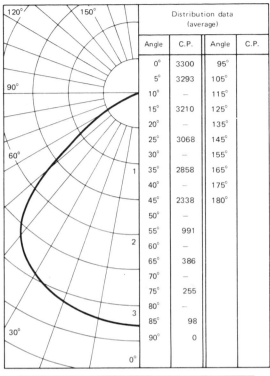

Distribution data (average)			
Angle	C.P.	Angle	C.P.
0°	3300	95°	
5°	3293	105°	
10°	—	115°	
15°	3210	125°	
20°	—	135°	
25°	3068	145°	
30°	—	155°	
35°	2858	165°	
40°	—	175°	
45°	2338	180°	
50°	—		
55°	991		
60°	—		
65°	386		
70°	—		
75°	255		
80°	—		
85°	98		
90°	0		

Output Data			
Zone degrees	Lumens	Percent total lamp lumens	Percent total distance
0–40°	14437	35.8	56.2
40–90°	3458	27.9	43.8
90–180°	0	0.0	0.0
0–180°	7895	63.7	100.0

COEFFICIENT OF UTILIZATION

Floor		For ρfc = 20%								
Ceiling		80%			50%			10%		
Walls		50%	30%	10%	50%	30%	10%	50%	30%	10%
RCR	1	0.68	0.66	0.64	0.64	0.63	0.61	0.59	0.58	0.57
	2	0.62	0.58	0.55	0.58	0.56	0.53	0.54	0.52	0.51
	3	0.55	0.51	0.48	0.53	0.49	0.46	0.50	0.47	0.45
	4	0.50	0.45	0.42	0.48	0.44	0.41	0.45	0.42	0.40
	5	0.46	0.40	0.36	0.43	0.39	0.36	0.41	0.38	0.35
	6	0.41	0.36	0.32	0.39	0.35	0.32	0.37	0.34	0.31
	7	0.37	0.32	0.28	0.35	0.31	0.28	0.34	0.30	0.27
	8	0.33	0.28	0.25	0.32	0.37	0.24	0.30	0.27	0.24
	9	0.30	0.25	0.21	0.29	0.24	0.21	0.27	0.24	0.21
	10	0.27	0.22	0.19	0.26	0.22	0.19	0.25	0.21	0.19

Maximum recommended spacing-to-mounting height ratio above work plane is: __1.43__

Figure 8-7. Typical Photometric Test Report

example problems

I. Longhand Power Calculation

Task Analysis[a]

General office: 100	E[b]	No. of tasks
Total task locations, 25		
T1. Reading fair reproductions, active filing	100	5
T2. Medium pencil or ink	70	16
T3. High-contrast printed forms	30	4
Accounting: 101		
Total task locations, 15		
T1. Auditing, bookkeeping, ink or medium pencil	150	10
T2. Intermittent filing	70	5
Private office: 102		
Total task locations, 2		
T1. Ink or medium pencil	70	2

[a]Information obtained from architect, client, or assumed.
[b]**IES Lighting Handbook,** 5th ed., Figure 9-80.

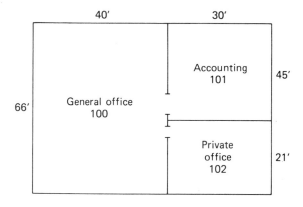

Power Calculation Solution, Longhand

Use luminaire LUMNO = DB117 (Figure 8-7). Luminaire meets efficiency criterion, $CU_{act} = 0.68$ 0.55 for tasks subject to veiling reflections.

1. General Office: 100

$$L = 66 \text{ ft}, \quad W = 40 \text{ ft}, \quad \text{gross area } A_G = 2640 \text{ ft}^2$$

$$h_{rc} = 5.5, \quad RCR = \frac{5(5.5)(66+40)}{66 \times 40} = 1.10, \quad CU = 0.68$$

Note: G stands for "gross," g stands for general.

T1. Five work locations at 100 fc.

$$A1 = \text{task area} = 5 \times 50 \text{ ft}^2 = 250 \text{ ft}^2$$

T2. Sixteen work locations at 70 fc.

$$A2 = \text{task area} = 16 \times 50 = 800 \text{ ft}^2$$

T3. Four work locations at 30 fc.

$$A3 = \text{task area} = 4 \times 50 = 200 \text{ ft}^2$$

Total task area:

Since, $\quad \Sigma A_T = 250 + 800 + 200 = 1250 \text{ ft}^2.$
$$\Sigma A_T = 1250 < \tfrac{1}{2}$$
$$A_G = 1320$$

$$\text{NTA} = A_G - EA_T = 2640 - 1250 = 1390 \text{ ft}^2$$

Therefore, the allowable area, *AA,* is 50 ft².

$$\text{watts} = \frac{fc \times A}{\text{lm/W} \times \text{CU}_{act} \times 0.70}$$

$$\text{watts}\,(A1) = \frac{100 \times 250}{55 \times 0.68 \times 0.70} = 955 \text{ W}$$

$$\text{watts}\,(A2) = \frac{70 \times 800}{55 \times 0.68 \times 0.70} = 2139 \text{ W}$$

$$\text{watts}\,(A3) = \frac{30 \times 200}{55 \times 0.68 \times 0.70} = 229 \text{ W}$$

General lighting level:

$$E_g = \frac{1}{3} \frac{(E1 \times A1) + (E2 \times A2) + (E3 \times A3)}{A1 + A2 + A3}$$

$$= \frac{1}{3} \frac{(100 \times 250) + (70 \times 800) + (30 \times 200)}{1250}$$

$$= 23.2 \text{ fc}$$

Nontask area, $\text{NTA} = A_G - \Sigma A_T = 2640 - 1250 = 1390 \text{ ft}^2$

Therefore,

$$A_g = \Sigma A_T = 1250 \text{ ft}^2$$

$$\text{watts } (A_g) = \frac{23.2 \times 1250}{55 \times 0.68 \times 0.70} = 1108 \text{ W}$$

Noncritical area,

$$\text{NCA} = \text{NTA} - A_g = 1390 - 1250 = 140 \text{ ft}^2$$

$$E_{NC} = \frac{1}{3} E_g = \frac{1}{3}(23.2) = 7.7 \text{ fc}$$

Therefore, use $E_{NC} = 10$ fc.

$$\text{watts } (A_{NC}) = \frac{10 \times 140}{55 \times 0.68 \times 0.70} = 53 \text{ W}$$

$$\text{total watts} = 955 + 2139 + 229 + 1108 + 53 = 4484 \text{ W}$$

2. Accounting: 101

$$L = 45, \qquad W = 30, \qquad A_G = 1350 \text{ ft}^2$$

$$h_{rc} = 5.5, \qquad \text{RCR} = 1.53, \qquad \text{CU} = 0.65$$

T1. Ten work locations at 150 fc.

$$A1 = 10 \times 50 = 500 \text{ ft}^2$$

T2. Five work locations at 70 fc.

$$A2 = 5 \times 50 = 250 \text{ ft}^2$$

$$\Sigma A_T = 500 + 250 = 750 \text{ ft}^2$$

Since, $\Sigma A_T = 750 > \frac{1}{2} A_g = 675$

Weighted task areas for allowable area:

$$AA = \frac{A_G}{2 \, (\Sigma \text{ no. loc.})} = \frac{1350}{2(10+5)} = 45 \text{ ft}^2$$

Revised task areas:

$$A1 = 10 \times 45 = 450 \text{ ft}^2$$

$$A2 = 5 \times 45 = 225 \text{ ft}^2$$

$$\text{watts } (A1) = \frac{150 \times 450}{55 \times 0.65 \times 0.70} = 2697 \text{ W}$$

$$\text{watts } (A2) = \frac{70 \times 225}{55 \times 0.65 \times 0.70} = 630 \text{ W}$$

for a program developed by the   41.0 fc

ft^2

al)

,   CU = 0.60

)3 W

· 100 = 530 ft^2

= 101 W

– 100 = 430 ft^2

$\dfrac{}{0}$ = 186 W

6 = 590 W

II. Computerized Power Calculation

The following is a sample terminal input/output
author and a sample of worksheet for data input.

```
PROJECT NAME
? XYZ CORP.
PROJECT NUMBER(I5)
? 75000
```

Data Input
(from a computer termi

```
 ROOM NUMBER(I3)
? 100
WIDTH
? 40
LENGTH
? 66
HEIGHT OF ROOM CAVITY
? 5.5
EFFICACY
? 55
 LUMINAIRE SEQ(I5) NO.-RIGHT ADJ.
? DB117
NUMBER OF TASKS(I1)
? 3
TASK 1 — E
? 100
TASK 1 — WORK     LOCATIONS
? 5.
TASK 2 — E
? 70.
TASK 2 — WORK     LOCATIONS
? 16.
TASK 3 — E
? 30.
TASK 3 — WORK     LOCATIONS
? 4.
```

Computer Output

ROOM NO. 100

TOTAL TASK AREA =	1250		
TOTAL GEN. AREA =	1250	GENERAL FC =	23
TOTAL NON-CRIT AREA =	140	NON-CRITICAL FC =	10
RCR =	1.10	CU =	.68

TASK WATTS = 3323
GENERAL WATTS = 1108
NON-CRIT. WATTS = 53

TOTAL ROOM WATTS = 4484
ROOM WATTS/SQ.FT. = 1.70

NUMBER OF IDENTICAL ROOMS = 1
TOTAL WATTS IDENTICAL ROOMS = 4484

TABLE 8-3

Power calculation worksheet

PROJECT NAME _Office Complex_   PROJECT NUMBER _7500_

CLIENT _XYZ Corp_   DATE _1980_   BY _RNH_   PAGE _1_ OF _1_

Room number (13)	Width	Length	HRC	Efficacy	Lum. seq. (13)	No. tasks	Task 1		Task 2		Task 3	
							E	Work location	E	Work location	E	Work location
100	40.	66.	5.5	55	DB117	3.	100.	5.	70.	16.	30.	4.
101	30.	45.	5.5	55	DB117	2.	150.	10.	70.	5.		
102	21.	30.	5.5	55	DB117	1.	70.	2.				

9

exterior lighting

Exterior lighting must be well planned to optimize the use of exterior facilities while minimizing energy consumption. Security and safety are the primary functional needs associated with most exterior applications. Roadway, area, and floodlighting are the three design techniques available to satisfy these functional needs.

Once a functional need has been established, the designer must select the most efficient equipment to maximize the utilization of the energy while minimizing power consumption. The quality of illumination will have a direct bearing on the safety, efficiency, and appearance of the system. Quality is dependent upon glare control, transient adaptation, color, and aesthetics.

Glare and transient adaptation will affect the safe movement of pedestrians and vehicles. Good glare control will minimize wasted spill light, which will increase the overall efficiency of the system by increasing the utilization of energy. Direct glare (Chap. 7) is the major problem encountered in most exterior lighting. It causes annoyance, discomfort, and/or a loss in visibility. The sensation of glare results from the luminance of a source being much higher than the surrounding luminance to which the eye is adapted. An emphasis on direct glare has been brought to the forefront in the past two years because of the awareness on the part of people of light pollution in the environment. Transient adaptation relates to the effect created by changing the direction of view and exposing the visual system to variations in the luminance patterns in the field of view. For example, as one drives down a roadway, the visual system adapts to an ambient level of luminance. Excessively bright light sources, such as oncoming headlights and streetlights, momentarily affect the adaptation state. The greater the luminance difference (luminance ratios) between the source and the surrounding area, the slower the rate at which the visual system will return to its original adaptation level. It is during this readaptation process that a loss in visibility is experienced that could result in an accident. A reduction in glare, and hence luminance ratios, will also result in a reduction in the amount of clutter and confusion reaching the visual system. This decrease in clutter and confusion will create a more pleasant, safer environment for pedestrians and vehicles.

Color is also important in the exterior environment since it influences the mood and behavior of people and thus their performance. High performance is essential to the security of facilities as well as safety.

Exterior lighting can be classified into static or movable systems. A static system is one in which the luminaire is not adjustable on its support and is positioned toward a single fixed point. Most roadway and pedestrian luminaires would be classified as static systems. A movable system is one in which the equipment can be aimed at different points. Floodlighting equipment with adjustable support mechanisms would be classified as movable.

Light sources are described in detail in Chap. 4. A brief summary of light sources as they apply to exterior lighting is included in this section.

1. Incandescent lamps should be avoided for exterior applications because of their low efficiency and short life. Incandescent lamps create high operating and maintenance costs.

2. Fluorescent lamps should be avoided for exterior applications because of their sensitivity to temperature and the poor quality of optical control. Enclosed and gasketed luminaires may maintain a sufficiently high ambient temperature to allow for operation under low temperatures; however, the enclosure that holds the heat in the unit during cold weather also holds the

heat in during warm weather, which causes a drop in the lumen output (Chap. 4) of the lamps. Because of the physical size of the light source, optical control will be poor, resulting in lower utilization characteristics.

3. Low-pressure sodium lamps produce monochromatic yellow light which turns all colors gray, brown, or black, except yellow. The lamp alone has a very high efficiency. However, when the source is combined with a ballast and luminaire, the overall efficiency of the system is low. Because of the physical size of the source, optical control is poor, resulting in low utilization characteristics.

4. Mercury vapor lamps require a phosphor coating if color rendition is to be acceptable. The phosphor-coated lamp is a large source, which means that optical control is poor and utilization decreases. The mercury vapor lamp also has a relatively low efficacy, which makes it the third choice for exterior applications.

5. Metal halide and high-pressure sodium lamps have relatively small light-emitting elements (arc tubes), which allow for good optical control. Each of the two sources has high lamp efficacy and good system efficiency. Thus these two sources are the top choices for exterior applications. The metal halide lamp has better overall color balance and is preferred where color is important. The high-pressure sodium lamp has a higher lamp efficacy but a dominant orange appearance that may be objectionable for some applications.

ROADWAY LIGHTING

A roadway may require lighting to facilitate the safe movement of people from one area to another. Roadway lighting is especially important where mixed modes of movement converge, such as vehicular/pedestrian and automobile/ service vehicle. These roadway systems should use low-glare luminaires that produce very little intensity between 70 and 90° from nadir and no intensity 90° or above. A reduction in intensity at these high angles (70 to 90°) will reduce light pollution, minimize spill light, and optimize the use of energy by placing the light where it is needed. High-angle intensity (70° and up) does little or nothing to produce illumination on a horizontal roadway surface.

The current roadway lighting standard[1] uses horizontal illumination and uniformity of illumination as the design criteria for roadway lighting. Both factors are deficient in terms of what the visual system actually sees. This is less than desirable, but it is currently the only approved procedure available. The designer should be aware of the deficiencies, but must use the standard until a new one becomes available.

Uniformity

The human visual system does not see the luminous energy coming *to* a surface (illumination); it sees the energy coming *from* a surface (luminance) (see

Figure 9-1). A luminance design procedure is much more complex and requires information on the reflectance properties of the roadway surface. However, with the aid of the computer, the problem is not as impossible as one might believe. Uniformity of illumination is also meaningless when talking about the uniform appearance of the pavement. The uniformity of the pavement is dependent upon the amount of light reflected toward the eyes (luminance) from various points on the surface. The pavement luminance is equal to the bidirectional reflectance distribution function (Chap. 3) times the horizontal illumination.

$$L = \beta_{(v_s, h)} E_h$$

where

L = luminance of the pavement surface
$\beta_{(v_s, h)}$ = bidirectional reflectance distribution function (**BRDF**)
E_h = illumination on a horizontal surface

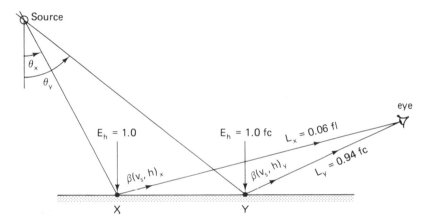

Figure 9-1. Uniformity: Illumination Versus Luminance

Assume that E_h = 1.0 fc; if $\beta_{(v_s, h)} X$ = 0.0567 for point X on the road, then,

$$L_X = 0.0567 \times 1.0 = 0.06 \text{ fL}$$

If $\beta_{(v_s, h)_Y}$ = 0.9409 for point Y on the road,

$$L_Y = 0.9409 \times 1.0 = 0.94 \text{ fL}$$

This represents almost a 16 to 1 uniformity ratio in terms of luminance (what the eye sees) for a 1 to 1 uniformity ratio of illumination.

Terminology (Figure 9-2)

1. Longitudinal roadway line (LRL): reference lines that run parallel to the curb.

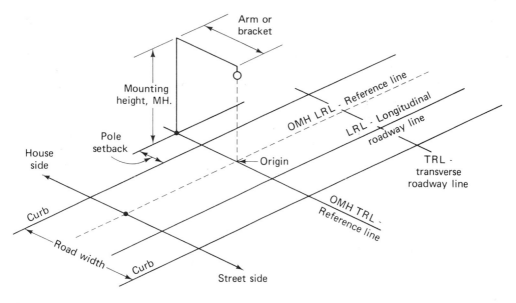

Figure 9-2. Terminology for Roadway Lighting

2. Transverse roadway line (TRL): reference lines that are perpendicular to the curb line or at 90° to the LRL.

3. Origin: directly under the luminaire and the point defined as the intersection of the zero mounting height (0MH) LRL and the 0MH TRL.

4. Reference Line: the longitudinal roadway line that passes through the origin (0MH LRL).

5. Street side (SS): the space located in front of the Reference Line.

6. House side (HS): the space located behind the Reference Line.

7. Longitudinal distance (LD): a distance measured in a direction parallel to the curb between transverse roadway lines.

8. Transverse distance (TD): a distance measured perpendicular to the curb between longitudinal roadway lines.

9. Elevation angle, θ: the vertical angle measured from nadir (see Figure 9-3) in a vertical plane.

10. Horizontal angle, α: the horizontal or azimuth angle between vertical planes measured in the horizontal plane of the roadway. The horizontal angle is measured from the 0° vertical plane (see Figure 9-3).

11. Maximum cone: defines the elevation angle, θ, at which the maximum candlepower (luminous intensity) occurs (see Figure 9-4).

12. Maximum vertical plane: the vertical plane in which the maximum candlepower (luminous intensity) occurs. The maximum vertical plane is defined in terms of the horizontal angle, α, from the 0° vertical plane to the vertical plane containing the maximum candlepower.

234

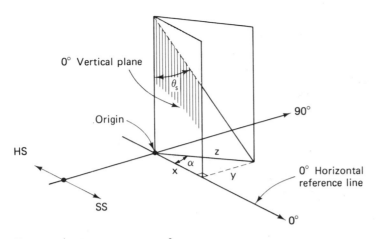

Figure 9-3. Elevation Angle, θ, and Horizontal Angle, α

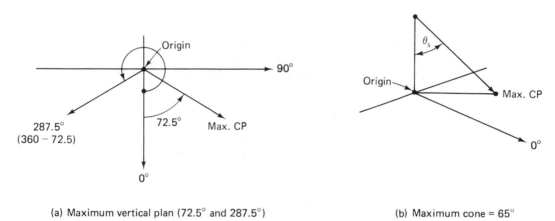

(a) Maximum vertical plan (72.5° and 287.5°) (b) Maximum cone = 65°

Figure 9-4. Maximum Vertical Plane and Maximum Cone

Roadway Luminaire Classification

The light distribution from roadway luminaires is classified in terms of the following:

1. Spread (short, medium, long), which describes the vertical light distribution.
2. Type (I, II, III, IV, V), which defines the lateral light distribution.
3. Control, which is divided into three categories (cutoff, semicutoff, non-cutoff).

Spread. The classification according to spread is determined by measurements in a longitudinal direction (see Figure 9-5). The spread is assigned on the basis of the location of the intersection of the maximum candlepower (luminous intensity) with the roadway surface.

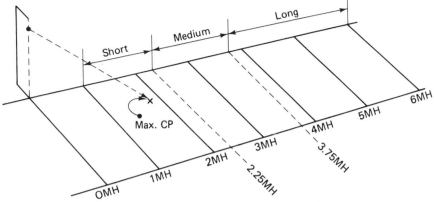

Short, medium or long—
Classification based on location of max. CP in a zone.

Figure 9-5. Spread Classification

Type. The type classification is determined by measurements in a transverse direction (see Figure 9-6). The type is assigned on the basis of the location of the half-maximum candlepower line that falls within the spread limits defined above.

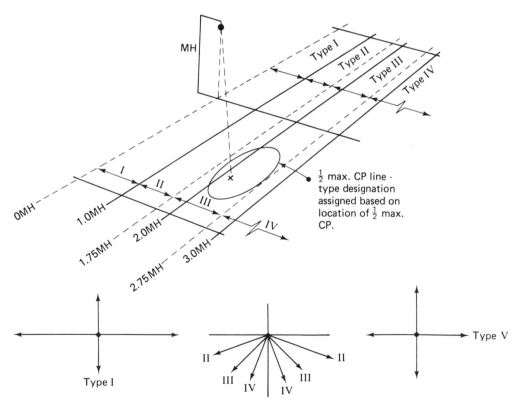

Figure 9-6. Type Classification

236

Control. Three descriptive terms are used to describe control: (1) cutoff, (2) semicutoff, and (3) noncutoff (see Figure 9-7). These terms are misleading and misrepresent the actual distribution characteristics of a luminaire (see Distribution Characteristics, this chapter).

Roadway Photometric Test Report

The standard photometric test report (Figure 9-22 in the data section) contains an isocandela diagram, isofootcandle curve, utilization curve, descriptive information, and a summation of distribution characteristics.

1. Cutoff

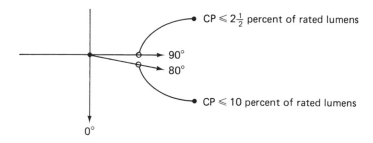

2. Semicutoff

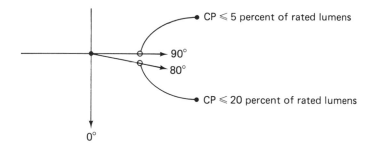

3. Noncutoff

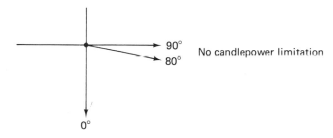

Figure 9-7. Control Classification

Isocandela Diagram

The isocandela diagram (Figure 9-23 in the data section) is usually plotted on a rectangular coordinate grid known as a **rectangular web.** The rectangular web has equally spaced vertical (longitudinal) lines and equally spaced horizontal (latitude) lines. Construction of the rectangular web from a sphere (Figure 9-8) is accomplished by straightening out the longitudinal line. The lines of latitude must increase in length to a value equal to the circumference of the equator while re-

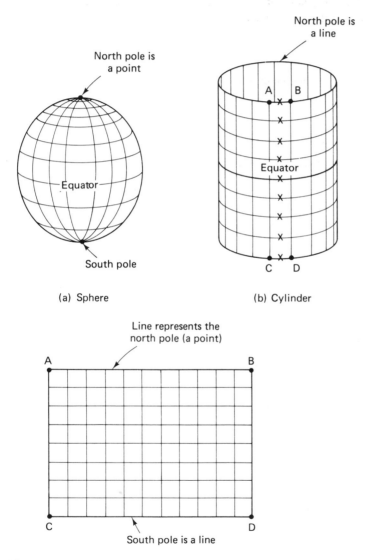

(a) Sphere (b) Cylinder

Figure 9-8. Rectangular Web Grid

238

maining parallel to the equator. This forms a cylinder of equally spaced longitudinal lines and lines of latitude. If the latitude lines are cut along a constant longitude, the cylinder can be flattened out to form the rectangular web.

The isocandle diagram plots the luminous intensity values for various angles of elevation and azimuth (I_θ, α) on a rectangular web. A vertical line passing through an azimuth angle represents a vertical plane of luminous intensity (candlepower; CP) that can be plotted on polar coordinate paper.

Transverse roadway lines are curved lines (Figure 9-9) that begin at $\alpha = 0°, \theta = 90°$ and end at $\alpha = 180°, \theta = 90°$. By starting at infinity (∞) and moving along a transverse roadway line (TRL) toward the luminaire, the angular change (α and θ) of the TRL can be calculated and plotted on the rectangular web. (see Figure 9-10).

HORIZONTAL ANGLE, α.

At $x = \infty$

$$\alpha = \arctan \frac{y}{\infty}$$

Therefore, $\alpha = 0°$. Moving toward the luminare. $x \to 0$; then

$$\alpha = \arctan \frac{y}{0},$$

therefore, $\alpha \to 90°$

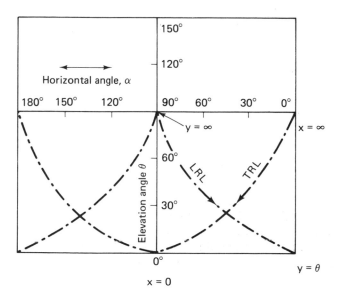

Figure 9-9. Isocandela Diagram: Transverse and Longitudinal Roadway Lines

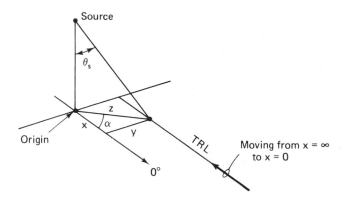

Figure 9-10. Transverse Roadway Line on Isocandela Diagram (Figure 9-9)

VERTICAL ANGLE, θ

As $x \to \infty$
$$\theta = \arctan \frac{\sqrt{(\infty)^2 + (y)^2}}{h}$$

Therefore, $\theta = 90°$

Moving toward the luminaire, $x \to 0$; then

$$\theta = \arctan \frac{\sqrt{(0)^2 + (y)^2}}{h}$$

Therefore, $\theta \to 0°$

SUMMARY.

At $x = \infty$,
$$\alpha = 0°, \qquad \theta = 90°$$

As $x \to 0$,
$$\alpha \to 90°, \qquad \theta \to 0°$$

Longitudinal roadway lines (Figure 9-9) are curved lines that begin at $\alpha = 90°$, $\theta = 90°$, and fan out to $\alpha = 0°$ ($\alpha = 180°$), $\theta = 0°$. Approaching the luminaire from infinity (∞) along a longitudinal roadway line, the angular change of the LRL can be calculated and plotted (see Figure 9-11).

HORIZONTAL ANGLE, α.

At $y = \infty$,
$$\alpha = \arctan \frac{\infty}{x}$$

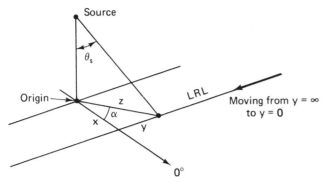

Figure 9-11. Longitudinal Roadway Line on Isocandela Diagram (Figure 9-9)

Therefore, $\alpha = 90°$

Moving toward the luminaire, $y \to 0°$; then

$$\alpha = \arctan \frac{\bar{0}}{x}$$

Therefore, $\alpha \to 0°$

VERTICAL ANGLE, θ.

At $y = \infty$,

$$\theta = \arctan \frac{\sqrt{(x)^2 + (\infty)^2}}{h}$$

Therefore,

$$\theta = 90°$$

Moving toward the luminaire, $y \to 0°$; then

$$\theta = \arctan \frac{\sqrt{(x)^2 + (0)^2}}{h}$$

Therefore, $\theta \to 0°$

SUMMARY.

At $y = \infty$,

$$\alpha = 90°, \qquad \theta = 90°$$

As $y \to 0°$,

$$\alpha \to 0°, \qquad \theta \to 0°$$

A simple example will demonstrate how the isocandela diagram can be used to find the luminous intensity arriving at a point on the roadway surface.

Example: What is the luminous intensity arriving at a point 40 ft in front of a luminaire and 25 ft down the roadway?

$$x = 40 \text{ ft}, \qquad y = 25 \text{ ft}, \qquad h = 30 \text{ ft}$$

$$\theta = \arctan \frac{\sqrt{x^2 + y^2}}{h}$$

$$= \arctan \frac{\sqrt{(40)^2 + (25)^2}}{30} = 57.5°$$

$$\alpha = \arctan \frac{y}{x} = \arctan \frac{25}{40} = 32.0°$$

Isofootcandle Diagram

The isofootcandle diagram (Figure 9-24 in the data section) is a contour map of illumination values. Illumination values of equal magnitude are connected together with lines called **iso** lines. The isofootcandle diagram can be used to calculate the illumination at a point (or points) on the roadway surface.

Two ratios must be calculated to determine the illumination at a point from the isofootcandle diagram: (1) transverse distance (TD) to mounting height (MH), and (2) longitudinal distance (LD) to mounting height (MH).

Example: What is the illumination at a point 40 ft in front of the luminaire and 25 ft down the road (see Figure 9-12) for a luminaire mounted 30 ft above the roadway?

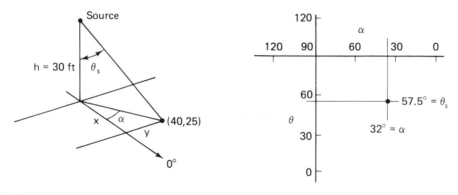

Figure 9-12. Use of Isocandela Diagram to Find $T(\theta, \alpha)$

$$x = 40 \text{ ft, transverse distance}$$
$$y = 25 \text{ ft, longitudinal distance}$$

Ratios

$$\frac{TD}{MH} = \frac{40}{30} = 1.33$$

$$\frac{LD}{MH} = \frac{25}{30} = 0.83$$

From Figure 9-13, the initial illumination at the point is 0.83 fc.

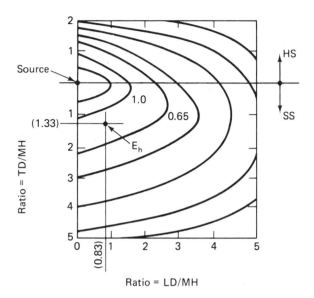

Figure 9-13. Use of Isofootcandle Diagram to Find E_h

Utilization Curve

The utilization curve indicates the amount of light falling on the roadway surface in terms of a coefficient. The utilization factor (coefficient of utilization) represents the percentage of rated bare lamp lumens that falls on two horizontal (one in front and one behind) strips of infinite length beneath the luminaire. It is a measure of the efficiency of a luminaire in terms of the luminaire's ability to direct light to the roadway surface. The utilization factor (UF) is used in the lumen formula to determine the average uniform horizontal illumination or to determine the spacing for some design level of illumination (see Figure 9-14).

Figure 9-15 shows utilization curves for five roadway luminaires. The differences can best be seen by studying the distribution characteristics of the luminaires (Figure 9-17). The utilization factor (UF) is found by entering the appropriate curve (street side or house side) with the corresponding ratio of transverse distance (TD) to mounting height (MH).

The house side ratio and street side ratio are calculated:

$$\text{HS ratio} = \frac{\text{TD}}{\text{MH}} = \frac{9.5}{30} = 0.32$$

$$\text{SS ratio} = \frac{\text{TD}}{\text{MH}} = \frac{38.5}{30} = 1.28$$

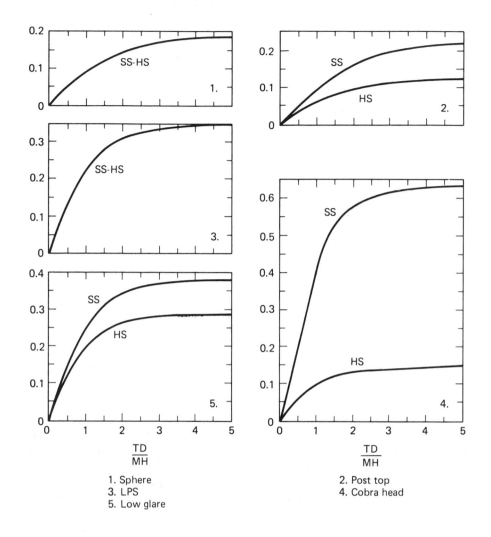

Figure 9-14. Utilization Curves for Five Luminaires

1. Sphere
3. LPS
5. Low glare

2. Post top
4. Cobra head

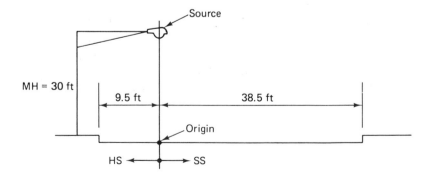

Figure 9-15. Geometry for Determination of Utilization Factor

244

From the utilization curves for luminaire 4 in Figure 9-14,

$$UF_{HS} = 0.05$$

$$UF_{SS} = 0.48$$

The total utilization factor is the sum of the house side and street side utilization factors:

$$UF_{TOT} = UF_{HS} + UF_{SS}$$

$$= 0.05 + 0.48 = 0.53$$

Distribution Characteristics

The distribution of luminous flux (lumens) from a luminaire can be used to rapidly evaluate a luminaire in terms of light pollution, wasted energy, glare, and efficiency. The luminous flux distribution of flux values represent the lumens distributed within a given zone (zonal lumens; see Figure 9-16). The standard practice is to divide a theoretical sphere surrounding the luminaire into four zones: (1) downward hemisphere house side (DW HS), (2) downward hemisphere street side (DW SS), (3) upward hemisphere house side (UP HS), and (4) upward hemisphere street side (UP SS). Since the luminaire is assumed to be placed inside of the sphere, all the luminous flux that is directed into the lower hemisphere, 0 to 90°, is summed to represent the downward component. All the luminous flux that is directed into the upper hemisphere, 90 to 180°, is summed to represent the upward component. The flux values for a typical roadway luminaire (luminaire 4,

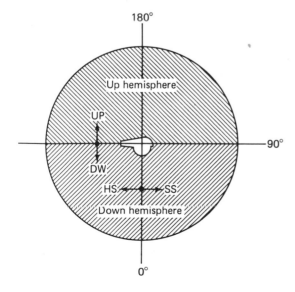

Figure 9-16. Distribution Terminology

Figure 9-14) are shown in Table 9-1 in accordance with standard practice for reporting the luminous flux distribution.

TABLE 9-1

Flux Values of Luminaire 4

Flux Values (%)

DW SS	63.6		
UP SS	1.2		
DW HS	14.5	Downward	78.1
UP SS	0.7	Upward	1.9
	80.0%	Total	80.0%

This luminaire has an efficiency of 80.0 percent. This means that 80.0 percent of the total bare lamp lumens actually gets out of the luminaire or 20.0 percent of the bare lamp lumens is absorbed or trapped inside the luminaire.

At first glance this may look quite good, but ask the following questions. What value is the 1.9 percent upward component, and will it ever reach the roadway surface? The 78.1 percent downward component looks good. It includes luminous flux in the entire downward hemisphere. What percentage of that luminous flux is at high angles (above 70°)? What effect does high-angle (70 to 90°) intensity have on glare? Is the cutoff control classification meaningful? How effective is high-angle intensity at producing horizontal illumination?

To answer these questions, the flux values must be reported for more zones in the downward hemisphere. Figure 9-17 shows the luminous flux distribution for the five luminaires (previously seen in Figure 9-14) in terms of output and percentage of total.

$$\% \text{ output} = \frac{\text{lumens in a zone}}{\text{bare lamp lumens}} \times 100$$

$$\% \text{ total} \quad = \frac{\text{lumens in a zone}}{\text{total lumens out (0-180°)}} \times 100$$

The sum of the % output for the house side plus street side from 0 to 180° represents the efficiency of the luminaire. The sum of the % total (HS + SS) from 0 to 180° will be 100 percent. A comparison of Figure 9-17 (luminaire) 4 with Table 9-1 will show the value of a more detailed breakdown of flux values.

Luminous Intensity Distribution

The optimum luminous intensity (candlepower) distribution to meet the standard design criterion (uniform horizontal illumination) can be constructed. The construction of the optimum distribution is based on the fundamental law

	1. Sphere				2. Post Top			
	% Output		% Total		% Output		% Total	
	SS	HS	SS	HS	SS	HS	SS	HS
0–40	4.1	4.1	4.3	4.3	3.2	2.4	6.4	4.8
40–70	9.3	9.3	9.9	9.9	11.5	6.2	23.2	12.6
70–90	8.4	8.4	9.0	9.0	10.1	5.9	20.4	11.9
90–180	25.1	25.1	26.8	26.8	5.4	4.9	10.9	9.9
0–180	46.8	46.8	50.0	50.0	30.2	19.4	60.9	39.1
	3. LPS				4. Cobra Head			
0–40	11.0	11.0	15.1	15.1	15.9	4.4	19.7	5.5
40–70	18.7	18.7	25.6	25.6	36.3	8.5	45.1	10.6
70–90	5.8	5.8	7.9	7.9	11.6	1.9	14.5	2.4
90–180	0.9	0.9	1.3	1.3	1.2	0.7	1.5	0.8
0–180	36.4	36.4	50.0	50.0	65.0	15.5	80.7	19.3
	5. Low Glare			
0–40	16.0	15.0	20.5	19.2
40–70	26.5	19.3	34.0	24.7
70–90	0.9	0.3	1.1	0.4
90–180	0.0	0.0	0.0	0.0
0–180	43.4	34.6	55.7	44.3

Figure 9-17. Flux Distribution for Five Luminaires

of illumination (inverse-square law) and the cosine law of illumination (see Figure 9-18). Since

$$E_h = \frac{I_\theta}{D^2} \cos \theta \quad \text{and} \quad \cos \theta = \frac{H}{D}$$

Therefore,

$$D = \frac{H}{\cos \theta}$$

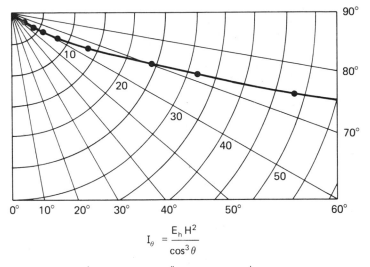

$$I_\theta = \frac{E_h H^2}{\cos^3 \theta}$$

Angle, θ	Multiplying factor	Angle, θ	Multiplying factor
0	1.0	60	8.00
10	1.05	65	13.25
20	1.21	70	24.99
30	1.54	75	57.68
40	2.22	80	190.98
50	3.77	85	1510.47
55	5.30	90	∞

$I_\theta = I_{0°} \times$ Multiplying factor @ θ — (to produce uniform horizontal illumination)

Figure 9-18. Candlepower Distribution for Uniform Horizontal Illumination

Then

$$D^2 = \frac{H^2}{\cos^2 \theta}$$

Therefore,

$$E_h = \frac{I_\theta}{H^2} \cos^3 \theta$$

To find the optimum luminous intensity, I_θ, for a constant E_h (uniform horizontal illumination), the equation can be rewritten as follows:

$$I_\theta = \frac{E_h H^2}{\cos^3 \theta}$$

If E_h = constant = 1 and H = 1 MH, then the luminous intensity, I_θ, must vary as the reciprocal of the cosine cubed.

$$I_\theta = \frac{1}{\cos^3 \theta}$$

Figure 9-18 shows that the luminous intensity at high angles (70 to 90°) must be magnitudes larger than the intensity produced at nadir (0°) for uniform horizontal illumination. At 70°, the luminous intensity must be 24.99 times the intensity at 0° (nadir), which is almost impossible to achieve. For reasons of economics, energy conservation, and glare control, the high angle (70 to 90°) intensity should be redirected into more useful zones in the distribution.

Roadway Lighting Equations

LUMEN FORMULA.   The lumen formula is used to calculate the average uniform illumination on the horizontal plane of the roadway or to calculate the spacing to achieve a required, maintained level of average uniform illumination (E_r). The initial illumination, E_i, is

$$E_i = \frac{\text{test lumens} \times \text{utilization factor}}{\text{pole spacing} \times \text{street width}}$$

The maintained illumination, E_m, is

$$E_m = E_i \times \text{light loss factor (LLF)}$$

Spacing, S, is

$$S = \frac{\text{test lumens} \times \text{UF} \times \text{LLF} \times \text{LF}}{E_r \times \text{street width}}$$

Once the spacing, S, has been calculated to satisfy the required horizontal illumination, the uniformity ratio specified in the standard practice[1] must be checked. The minimum-to-average uniformity of illumination is calculated by finding the minimum illumination level of the roadway. A layout configuration must be selected from the three standard layouts (Figure 9-19). The minimum illumination at one of four (or more) points must be determined. If the ratio of the minimum-to-average exceeds the recommended uniformity ratio, the design must be modified. Any one or all of the following variables can be changed to meet the uniformity ratio criterion: (1) spacing, (2) mounting height, (3) luminaire lateral placement, and/or (4) luminaire distribution. An example problem utilizing this design procedure can be found at the end of this chapter.

LIGHT LOSS FACTOR (LLF).   The LLF is a number less than unity that takes into account losses in light output due to dirt accumulation and lamp lumen depreciation.

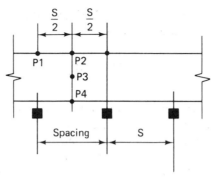

One side spacing

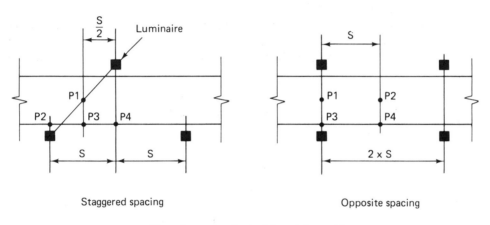

Staggered spacing   Opposite spacing

(P– points to be checked for minimum E)

Figure 9-19.   Spacing Configurations

LAMP FACTOR (LF).   The LF is used to increase the test lumens (lumens found on the test report) to the current catalog listed lumens, and to modify isofootcandles, isocandelas (luminous intensity or candlepower), and lumens (luminous flux).

$$LF = \frac{\text{catalog lamp lumens}}{\text{test lumens}}$$

MOUNTING HEIGHT CORRECTION (MHC).   The MHC is used to modify the footcandle values for mounting heights that are different from the test mounting height.

$$MHC = \left(\frac{\text{test mounting height}}{\text{actual mounting height}}\right)^2$$

UNIFORMITY RATIO.   The uniformity ratio is defined as a ratio of average illumination to minimum illumination on the roadway. The average-to-minimum ratio[1] should not exceed 3 to 1 except for residential areas, which should not exceed 6 to 1. The usefulness of this ratio was discussed previously in this chapter. (See also the example problem at the end of this chapter.)

$$\frac{E_{ave}}{E_{min}} \leqslant 3.0 \quad \text{(6.0 for residential areas)}$$

Visibility and Visual Comfort

Seeing objects on the roadway and proper guidance down the roadway are a function of driver alertness, visual comfort, and visibility. Visual comfort is affected by luminance ratios, fluctuation of luminance patterns, and field luminance. Visibility is affected by veiling luminances within the eye (reduces contrast), target (or object) luminance, pavement luminance, and the overall field luminance. Three complex interactions affect visual comfort and visibility: (1) luminaire/driver, (2) pavement/driver, and (3) luminaire/pavement.

1. Luminaire/driver: the luminaire affects the driver's visual comfort and visibility in terms of the luminaire's effect on adaptation, veiling luminance, sensitivity, transient adaptation, and luminance ratios.
2. Pavement/driver: the pavement affects the driver's visual comfort and visibility. Visual comfort is a function of the luminance ratios produced on the pavement and the uniformity of luminance. Nonuniform pavement luminances can create confusion and clutter that increase discomfort. Visibility is a function of the adaptation level, which is related to pavement luminance. Uniformity of pavement luminance and the level of pavement luminance can affect the visibility of objects on the roadway in terms of the luminance difference between the pavement and object.
3. Luminaire/pavement: the distribution characteristic of the luminaire and the reflectance properties (BRDF) of the pavement surface affect uniformity and level of luminance. Luminaire distribution can also increase object visibility in terms of vertical surface illumination, which can increase depth perception and modeling of the object for easy identification.

Disability Veiling Brightness (DVB) (Figure 9-20)

High DVB can reduce object visibility and visual acuity by reducing object contrast in space as a result of veiling luminances produced in the eye. However, increasing DVB can increase contrast sensitivity, which means that object contrast can decrease. A driver location is selected and the individual DVB values for each luminaire in the visual field are calculated by the following formula. Windshield cutoff must be taken into consideration. The total DVB is the sum of the individual DVB values.

$$\text{DVB} = \frac{10\pi E_v}{\theta_G^2}$$

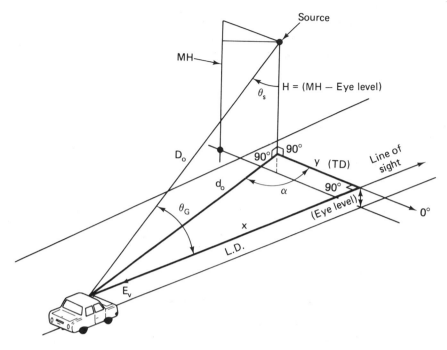

Figure 9-20. Disability Veiling Brightness Terminology

where E_v = vertical illumination at the eye on a plane normal (perpendicular) to the line of sight

θ_G = angle between a line from the luminaire to the eye and the line of sight

$$E_v = \frac{I_{(\theta_s, \alpha)} \cos \theta_G}{D_o^2}$$

$$\cos \theta_G = \frac{x}{D_o} = \frac{LD}{D_o}$$

Therefore,

$$E_v = \frac{I_{(\theta_s, \alpha)} \times LD}{D_o^3}$$

Then for a DVB/1000 cd [assume that $I_{(\theta_s, \alpha)} = 1000$ cd]

$$DVB = \frac{10\pi (1000) \times LD}{D_\theta^3 \theta_G^2}$$

where

$$D_o = \sqrt{(d_o)^2 + (H)^2}$$

$$d_o = \sqrt{(x)^2 + (y)^2} = \sqrt{(LD)^2 + (TD)^2}$$

H is the vertical distance from eye level to the luminaire

$$\theta_G = \text{arccosine } \frac{x}{D_o} = \text{arccosine } \frac{LD}{D_o}$$

To find $I_{(\theta_s, \alpha)}$, where θ_s is the elevation angle from nadir,

$$\theta_s = \arctan \frac{d_o}{H}$$

$$\alpha = \arctan \frac{x}{y} = \arctan \frac{LD}{TD}$$

AREA LIGHTING

Area lighting may be required to provide security lighting and for the safe movement of people. Area lighting can increase security by increasing visibility, thus reducing crimes against property. The safety aspects of area lighting not only include the prevention of crimes against people but also reduction in accidents.

The use of movable or aimable systems for area lighting is quite common. The approach has been to place floodlighting units on poles and aim them out to flood an area with light. One must remember that, as a unit is tilted up and out, the internal parts of the luminaire and light source are exposed to the direct line of sight. This approach may be initially less expensive, but it creates glare, excessive spill, and wasteful energy utilization. Because of the effect on transient adaptation, the visibility may be reduced to a point that is detrimental to both security and safety.

The trend in area lighting has been to move away from movable or aimable systems to the static systems used in roadway lighting. The use of low-glare, high-efficiency systems is recommended for area light. The same principles and procedures outlined in the roadway lighting section of this chapter apply to area lighting.

BUILDING LIGHTING

Building lighting can increase the aesthetic appearance of a selected structure as well as act as security lighting. Lighting of the exterior surface of the building can reduce crimes such as burglary and theft by making the exterior surfaces more visible to security personnel.

The primary concern in building floodlighting is to minimize glare and maximize the distribution of illumination on the building surface. The required level of illumination is a function of the reflective properties of the building, the ambient surrounding luminances, and the aesthetic appearance desired. Building floodlighting is the terminology used to describe ground-mounted or pole-mounted equipment that is set back some distance from the building. Building lighting can also be achieved by equipment mounted directly on the structure.

The type of area to be lighted, the location of equipment, and the variation in surrounding conditions impose design problems that make standardization difficult if not impossible.

The lighting of commercial buildings can serve as a means of advertising. In other words, let the architecture sell the product by attracting attention to the building and by creating a favorable impression on passersby. If the building is lighted properly, there should be no need for additional advertising to distract from the architecture.

Floodlighting is dependent primarily on the ability of the designer to manipulate brightness, texture, shadow, and color. The designer must establish the architectural effect that is desired with the floodlighting. It is helpful to study the daylight effects on the building. This allows the designer to select the strong daylight effects that are desirable and to attempt to reproduce these effects under floodlighting.

Daylighting effects are difficult to produce because of the directional characteristics of the sun and the diffuse characteristics of the sky. The sun is warm in color, while the diffused sky light is cool. Shadows are never black but less bright and bluer. Daylight is not constant but continually varying from day to day and hour to hour.

Illumination Levels

The illumination levels given in the **IES Lighting Handbook**[2] are recommended values that may vary depending on the design effects being created.

Type of Equipment

Floodlighting equipment can be supplied as either open or enclosed equipment. Enclosed equipment has a higher light-loss factor and usually more accurate beam control. Open units usually require special bulbs with hard glass to prevent breakage caused by cold elements touching the hot bulb. Enclosed equipment is initially more expensive; however, over a long period of time the savings due to higher light loss factors and the use of standard lamps will usually pay for the higher initial cost.

Terminology (Figure 9-21)

1. Maximum candlepower (luminous intensity): the maximum candlepower value recorded in the beam pattern; for example, the maximum candlepower is 47,775 cd.

2. Beam limit: that part of the beam in which the candlepower values are 10 percent or more of the maximum candlepower.

$$\text{limiting CP line} = 10\% \times \text{max. CP}$$

For example,

$$10\% \times 47,775 = 4778 \text{ cd} = \text{beam limit}$$

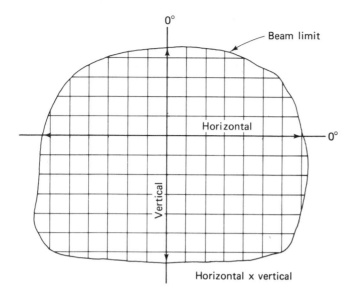

Figure 9-21. Definitions for Floodlighting

3. Beam lumens (BL): includes only the lumens within the beam limit. For example,

$$\text{beam lumens} = \Sigma \text{ lumens within the beam limit}$$

4. Coefficient of beam utilization (CBU) or utilization factor (UF): the ratio of the lumens effectively lighting an area to the total available beam lumens expressed in a percentage

$$\text{CBU} = \text{UF} = \frac{\Sigma \text{ lumen within beam limit falling on bldg}}{\Sigma \text{ beam lumens}}$$

The CBU is dependent upon the following:

a. Equipment location.

b. Aiming point.

c. Lumens within the beam limit.

The average CBU usually will fall within the range of 0.60 to 0.90. If less than 0.60 of the beam lumens are utilizcd, the solution will be uneconomical and a different location or narrower beam floodlight should be used. If the CBU is over 0.90, the beam spread is too narrow and the result will be spotty. A typical design begins by assuming that CBU = 0.70.

5. Beam spread: the angle, measured in degrees, between the beam limits both horizontally and vertically (Figure 9-21).

6. Beam classifications:

Beam spread (degrees)	NEMA type
10–18	1
18–29	2
29–46	3
46–70	4
70–100	5
100–130	6
130–up	7

For example,

$$\text{horizontal} \times \text{vertical}$$
$$75° \times 35°$$
$$\text{NEMA type } 5 \times 3$$

7. Beam efficiency: ratio of total lumens in the beam limit to lumens generated by the bare lamp.

$$\text{beam efficiency} = \frac{\text{total lumens in beam limit}}{\text{total bare lamp lumens}}$$

8. Total efficiency:

$$\text{total efficiency} = \frac{\text{total lumens out of the luminaire}}{\text{total bare lamp lumens}}$$

9. Beam footcandle: terminology used by some manufacturers to specify footcandles in terms of normal footcandles, **not** horizontal footcandles.

Floodlight Photometric Test Report

The standard floodlight photometric test report (Figure 9-25 in the data section) consists of three major components: (1) equipment description and tabulated information, (2) isocandela curves, and (3) beam lumens within the test grid. The isocandela curves and lumens are presented on a rectangular coordinate grid. The isocandela plot is placed on half of the rectangular grid system. Lumens for each rectangular areas are calculated and placed with the rectangular areas on the other half of the rectangular coordinate diagram. The CBU is calculated (see example problem at the end of this chapter) by superimposing the building projection onto the rectangular coordinate system. The lumens falling on the projection of the building can be summed, and the CBU calculated for the particular luminaire and building geometry. With this type of coordinate system, the top and bottom of the building will be horizontal lines, while the sides will be slightly curved.

Design Procedure

1. Determine the level of illumination.
2. Determine the type and location of equipment.
3. Estimate the light loss factor.
4. Determine the number of floodlights required.
5. Determine the actual coefficient of beam utilization (CBU).
6. Check for coverage and uniformity.
7. Readjust type or location of equipment if step 6 is inadequate.

Building Floodlighting Equations

Floodlight calculations consist of analyzing the coverage and uniformity of illumination produced on a building by (1) a lumen method or (2) point-by-point calculations.

Lumen Method

The lumen method is used to calculate the average uniformly maintained illumination on a building surface or to calculate the number of luminaires to produce a required level of illumination.

$$E_{ave} = \frac{N(BL)\,(CBU)\,(LLF)\,(OV)\,(TF)\,(LF)\,(LP)}{A}$$

where N = number of luminaires
BL = beam lumens of each luminaire
CBU = UF (utilization of beam lumens)
LLF = light-loss factor
OV = overvoltage factor. Operating at overvoltage is common in sports lighting; it results in increased light output and decreased life.
TF = temperature factor (usually only fluorescent)
LF = lamp factor $\dfrac{\text{catalogue lamp lumens}}{\text{test lamp lumens}}$
LP = lamp position factor, a function of the actual burning position; HID lamps burned in any position other than the position for which they were rated will have a change in output

For design purposes, the lumen formula is rearranged to

$$N = \frac{E_r \times \text{area}}{BL \times CBU \times LLF \times OV \times TF \times LF \times LP}$$

Point-by-Point Method

The point-by-point method (Chap. 7) utilizes the inverse-square law and the cosine law of illumination to calculate the illumination levels at various points on the building surface. This procedure lends itself to computerized design techniques. It is also used to check the uniformity of illumination on the building surface.

references

1. "American National Standard Practice for Roadway Lighting," ANSI/IES RP-8, Iluminating Engineering Society of North America, New York, 1977.

2. Kaufman, John, **IES Lighting Handbook,** 5th Ed. Illuminating Engineering Society, New York, 1972.

4.79

Vertical Angles

50
60
65
70
72.5
75
77.5
0
0

E
on.

		33	34	35
8	.83	.78	.74

ription:

A

etal Halide

TOFF TYPE III

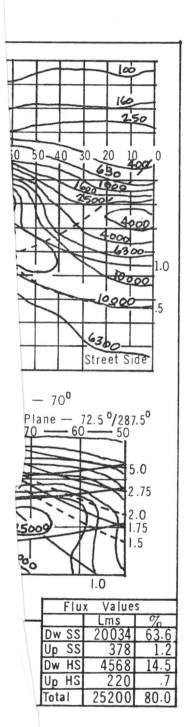

Street Side

— 70°

Plane — 72.5°/287.5°

Flux	Values	
	Lms	%
Dw SS	20034	63.6
Up SS	378	1.2
Dw HS	4568	14.5
Up HS	220	.7
Total	25200	80.0

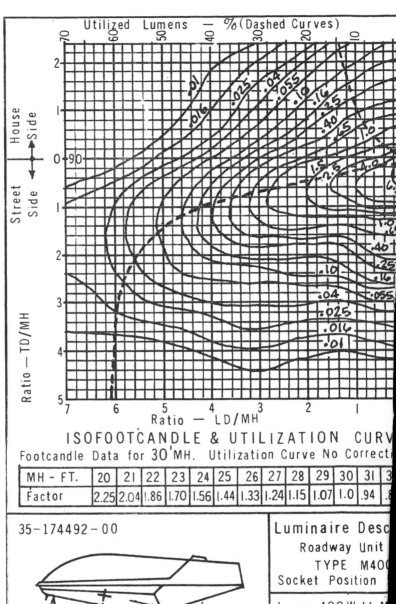

Utilized Lumens — % (Dashed Curves)

Ratio — TD/MH

House Side / Street Side

Ratio — LD/MH

ISOFOOTCANDLE & UTILIZATION CURV

Footcandle Data for 30' MH. Utilization Curve No Correcti

MH – FT.	20	21	22	23	24	25	26	27	28	29	30	31	3
Factor	2.25	2.04	1.86	1.70	1.56	1.44	1.33	1.24	1.15	1.07	1.0	.94	.8

35–174492–00

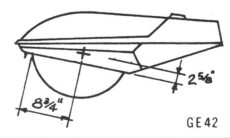

8¾"   2⅝"

GE 42

Luminaire Desc
Roadway Unit
TYPE M40
Socket Position

Lamp: 400 Watt M
Refractor 510
MEDIUM SEMI-CU

No. 4492

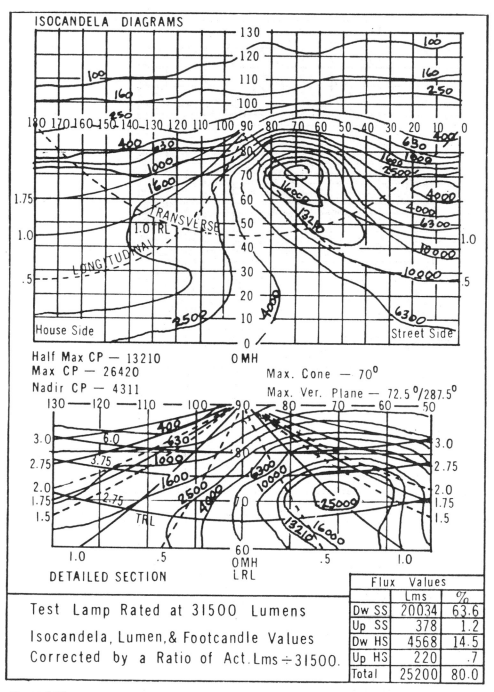

Half Max CP — 13210
Max CP — 26420
Nadir CP — 4311

Max. Cone — 70°
Max. Ver. Plane — 72.5°/287.5°

Test Lamp Rated at 31500 Lumens

Isocandela, Lumen, & Footcandle Values

Corrected by a Ratio of Act. Lms ÷ 31500.

Flux	Values	
	Lms	%
Dw SS	20034	63.6
Up SS	378	1.2
Dw HS	4568	14.5
Up HS	220	.7
Total	25200	80.0

Figure 9-23.

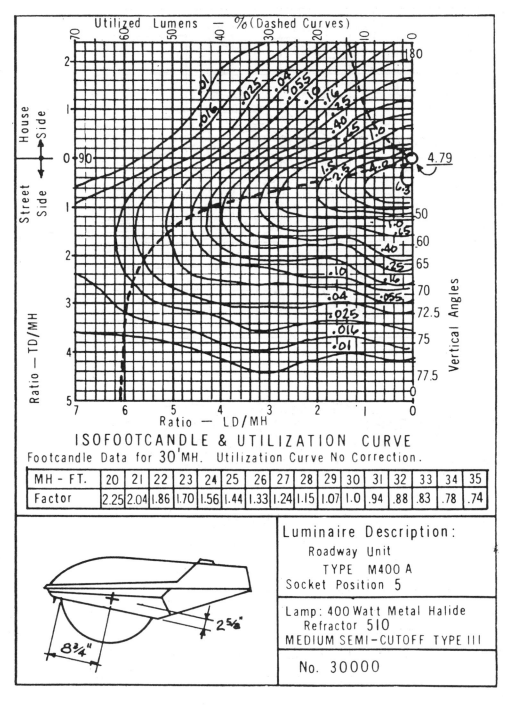

ISOFOOTCANDLE & UTILIZATION CURVE
Footcandle Data for 30' MH. Utilization Curve No Correction.

MH - FT.	20	21	22	23	24	25	26	27	28	29	30	31	32	33	34	35
Factor	2.25	2.04	1.86	1.70	1.56	1.44	1.33	1.24	1.15	1.07	1.0	.94	.88	.83	.78	.74

Luminaire Description:
Roadway Unit
TYPE M400 A
Socket Position 5

Lamp: 400 Watt Metal Halide
Refractor 510
MEDIUM SEMI-CUTOFF TYPE III

No. 30000

Figure 9-24. Isofootcandle Diagram

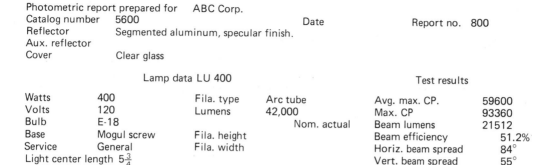

Photometric report prepared for   ABC Corp.
Catalog number   5600   Date   Report no.   800
Reflector   Segmented aluminum, specular finish.
Aux. reflector
Cover   Clear glass

Lamp data LU 400   Test results

Watts	400	Fila. type	Arc tube	Avg. max. CP.	59600
Volts	120	Lumens	42,000	Max. CP	93360
Bulb	E-18		Nom. actual	Beam lumens	21512
Base	Mogul screw	Fila. height		Beam efficiency	51.2%
Service	General	Fila. width		Horiz. beam spread	84°
Light center length $5\frac{3}{4}$				Vert. beam spread	55°

Test procedure and data form in accordance with I.E.S. and N.E.M.A. standards

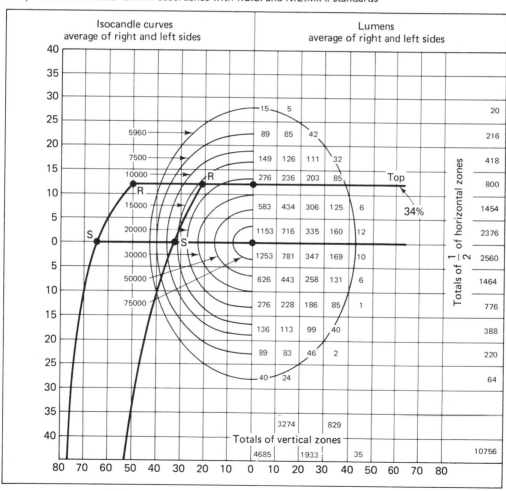

Figure 9-25.   Floodlight Photometric Test Report

example problems

EXAMPLE PROBLEM: ROADWAY LIGHTING

Given the following roadway geometry, determine the spacing required to produce 0.8 fc average maintained. Find the minimum footcandles and check the minimum-to-average illumination uniformity ratio. Use the luminaire shown in Figure 9-22 with staggered spacing.

Assume the following:

$$\text{catalog lumens} = 32{,}000 \text{ lm}$$
$$\text{MH} = 45 \text{ ft}$$
$$\text{LLF} = 0.70$$
$$E_r = 0.8 \text{ fc}$$

Cross section:

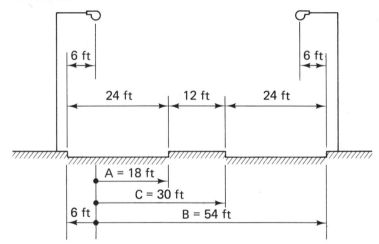

Part A: Spacing

$$\text{spacing} = \frac{\text{LMS} \times \text{UF} \times \text{LF} \times \text{LLF}}{E_r \times W}$$

Street Side

$$\frac{\text{TD}}{\text{MH}} = \frac{18}{45} = 0.40 \qquad\qquad \text{UF}_A = 0.16$$

$$\frac{\text{TD}}{\text{MH}} = \frac{54}{45} = 1.2, \qquad \text{UF}_B = 0.46 \left.\vphantom{\begin{array}{c}a\\b\end{array}}\right\} \qquad \text{UF}_{BC} = \text{UF}_B - \text{UF}_C = 0.16$$

$$\frac{\text{TD}}{\text{MH}} = \frac{30}{45} = 0.67, \qquad \text{UF}_C = 0.30 \qquad\qquad\qquad \text{UF}_{SS} = 0.32$$

264

House Side

$$\frac{TD}{MH} = \frac{6}{45} = 0.13 \qquad\qquad UF_{HS} = 0.02$$

$$\overline{UF_{TOT} = 0.34}$$

$$S = \frac{31,500 \times 0.34 \times \left(\frac{32.0}{31.5}\right) \times 0.70}{0.8 \times 48} = 198.3 \text{ ft} \quad \text{Use 200 ft oc.}$$

Part B: Uniformity Check

1. Find E_i at each point.

	$\dfrac{TD}{MH}$	$\dfrac{LD}{MH}$	E_i
Point 1			
Pole A	$\frac{54}{45} = 1.2$	$\frac{200}{45} = 4.44$	0.13
Pole B*	$\frac{6}{45} = 0.13$	$\frac{0}{45} = 0$	4.00
Pole C	$\frac{54}{45} = 1.2$	$\frac{200}{45} = 4.44$	$\underline{0.13}$
		$E_{p1_i} =$	4.26 fc
Point 2			
A *	$\frac{6}{45} = 0.13$	$\frac{200}{45} = 4.44$	0.025
B	$\frac{54}{45} = 1.2$	$\frac{0}{45} = 0$	1.25
C*	$\frac{6}{45} = 0.13$	$\frac{200}{45} = 4.44$	$\underline{0.025}$
		$\boxed{E_{p2_i} = 1.30 \text{ fc}}$	
Point 3			
A *	$\frac{6}{45} = 0.13$	$\frac{300}{45} = 6.67$	0
B	$\frac{54}{45} = 1.2$	$\frac{100}{45} = 2.22$	1.0
C *	$\frac{6}{45} = 0.13$	$\frac{100}{45} = 2.22$	$\underline{0.52}$
		$E_{p3_i} =$	1.52 fc
Point 4			
A	$\frac{18}{45} = 0.40$	$\frac{300}{45} = 6.67$	0.01
B	$\frac{30}{45} = 0.67$	$\frac{100}{45} = 2.22$	1.35
C	$\frac{18}{45} = 0.40$	$\frac{100}{45} = 2.22$	$\underline{1.25}$
		$E_{p4_i} =$	2.61 fc

*indicates that distances and illumination values are House
Side (HS).

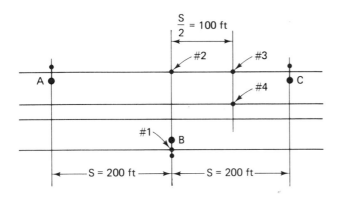

$$E_{min_i} = E_{p2_i} = 1.30 \text{ fc}$$

$$E_{min} = E_{min_i} \times \left(\frac{\text{test dist.}}{\text{act. dist.}}\right)^2 \times LF \times LLF$$

$$= 1.30 \times \left(\frac{30}{45}\right)^2 \times \left(\frac{32.0}{31.5}\right) \times 0.70 = 0.40 \text{ fc}$$

2. Check uniformity.

$$\frac{E_{ave}}{E_{min}} \leqslant 3.0, \text{ criteria,} \qquad \frac{E_{ave}}{E_{min}} = \frac{0.80}{0.40} = 2.00 < 3.00 \text{ (O.K.)}$$

Example Problem 2: Building Floodlighting

A light-white building located in a dark ambient surround is to be illuminated with ground-mounted floodlights set back 20 ft. Determine the actual CBU and check the uniformity. Use the floodlight photometric found in Figure 9-25 in the data section.

$E_r = 5 \text{ fc}$ area $= 75 \times 240 = 18,000 \text{ ft}^2$ (one side)

$LLF = 0.75$ assume CBU $= 0.70$

           BL $= 21,512$ (photometrics from Figure 9-25)

Solution

1. Calculate the number of floodlights required.

$$N = \frac{E \times A}{BL \times CBU \times LLF}$$

$$= \frac{5 \times 18,000}{21,512 \times 0.70 \times 0.75} = 7.97, \qquad \text{use 8}$$

Two luminaires per ground position will be tried for a total of four ground positions. Four of the eight luminaires (Figure 9-25) will be aimed at points 40 ft above ground. The remaining four luminaires will require a lower-wattage lamp, wider distribution, and lower aiming point. For the purposes of this example, only the top four luminaires will be analyzed. The maintained illumination at points *A, O, B, x,* and *y* will be checked for uniformity. The illumination contribution from the remaining four lower-wattage luminaires should be added to the points. For this example, they will not be included.

2. Determine the actual CBU.

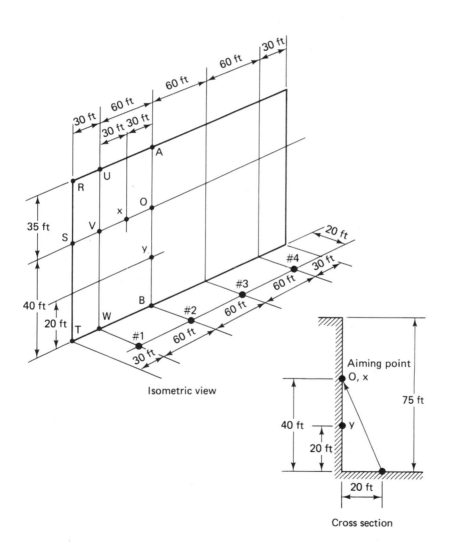

Isometric view

Cross section

Luminaire 1 (4)

1. Locate building on isodiagram for luminaire 1 (4).

$$\angle U1W = \arctan \frac{75}{20} = 75.1°$$

$$\text{aiming pt} < V1W \ \arctan \frac{40}{20} = 63.4°$$

$$\text{top} = 75.1° - 63.4° = 11.7°$$

$$\text{bottom} = 63.4°$$

$$\overline{U1} = \sqrt{75^2 + 20^2} = 77.6$$

$$\overline{V1} = \sqrt{40^2 + 20^2} = 44.7$$

Corner R from U:

$$\angle R1U = \arctan \frac{30}{77.6} = 21.1°$$

Edge S from V:

$$\angle S1V = \arctan \frac{30}{44.7} = 33.9°$$

Bottom corner T from W:

$$\angle T1W = \arctan \frac{30}{20} = 56.3°$$

2. Sum lumens on the building (columns from left to right). See Figure 9-25.

24	22	3274	4685	1933	829	35	
85	214	−5	−15	−42	−32		
104	335	−85	−89	−111	−29		
85	347	−126	−149	−124			
40	258	−156	−182				
2	186						
	99						
	46						
		2902	4250				
		× 2	× 2				sum
340	1506	5804	8500	1656	768	35	= 18,608

3. Calculate CBU for luminaire 1 (and 4).

$$CBU = \frac{18,608}{21,512} = 0.87$$

Luminaire 2

1. Locate building on isodiagram for luminaire 2 (3).
   Corner R from A:

$$\angle R2A = \arctan \frac{90}{77.6} = 49.2°$$

Edge S from O:

$$\angle S20 = \arctan \frac{90}{44.7} = 63.6°$$

Bottom corner T from B:

$$\angle T2B = \arctan \frac{90}{20} = 77.5°$$

2. Sum lumens on the building (columns). See Figure 9-28.

4685	3274	1933	829	35		
− 15	− 5	− 42	− 32			
− 89	− 85	− 111	− 29			
− 149	− 126	− 124				
− 182	− 156				½ sum	sum
4250	2902	1656	768	35 =	9611 × 2 =	19,222

3. Calculate CBU for luminaire 2 (and 3).

$$CBU = \frac{19,222}{21,512} = 0.89$$

The **actual CBU** is

$$CBU = \frac{(2 \times 0.87) + (2 \times 0.89)}{4} = 0.88$$

$$\text{actual CBU} = 0.88 > \text{assumed CBU} = 0.70$$

Assumption close enough to continue example problem. (Actual CBU is high, which indicates that the solution is probably going to be spotty and nonuniform. A change in aiming point of luminaire distribution might be warranted.)

3. Uniformity Check (cont.)

For luminaire 2, the contribution at points A, O, B, x, and y is as follows:

POINT A

$$E_A = \frac{24,000}{(77.6)^2} \cos 75.1 = 1.02 \times 0.75 = 0.77 \text{ fc}$$

POINT O

$$E_O = \frac{93,360}{(44.7)^2} \cos 63.4 = 20.92 \times 0.75 = 15.69 \text{ fc}$$

POINT y

$$L y 2B = \arctan \frac{20}{20} = 45°$$

$$\overline{y2} = \sqrt{20^2 + 20^2} = 28.3$$

The angle from aiming point O down to point y is

$$\angle O2B - Ly2B = 63.4° - 45° = 18.4°$$

$$E_y = \frac{9000}{(28.3)^2} \cos 45° = 7.95 \times 0.75 = 5.96 \text{ fc}$$

POINT x

$$\angle x2O = \arctan \frac{30}{44.7} = 33.9°$$

$$x2 = \sqrt{30^2 + (44.7)^2} = 53.8 \text{ ft}$$

$$E_x = \frac{14{,}000}{(53.8)^2} \cos 33.9° \cos 63.4° = 1.79 \times 0.75$$

$$= 1.34 \text{ fc}$$

POINT B

The angle from the aiming point O down to the bottom of the building B is

$$\angle O2B = 63.4°$$

The bottom of the building is off the diagram; therefore,

$$E_B = 0 \text{ fc}$$

This portion of the building would be illuminated by the four lower-wattage, wide-distribution luminaires (not analyzed in this example).

4. Conclusions

Concerning maintained illumination produced by luminaire 2 (only) at points A, O, B, x, and y, it can be concluded that the uniformity ratio between the top (A) and the aiming point (O) exceeds 20 to 1, which is excessive. Examination of Figure 9-25 will show that luminaires 1 and 3 will not add significantly to point (A). Therefore, the aiming point should be moved up the building or a new luminaire should be selected. Once the layout has been modified, the maintained illumination levels must be recalculated as shown above.

10 daylighting design

This chapter attempts to take an objective look at the complex interaction and impact of daylighting on the human occupants, luminous environment, and energy conservation. Daylighting is dealt with by first analyzing daylighting and then establishing design techniques. The analysis is in terms of the three primary aspects of daylighting: (1) exterior environment, (2) interior environment, and (3) interface medium (the window). The analysis of these components should be used to establish design parameters and techniques.

ANALYSIS OF DAYLIGHTING

The primary difficulty in daylighting is the variability of daylight with respect to the time of day and year, location (i.e., with geographical latitude), and environmental conditions. These variations in the quality and quantity of daylight (exterior environment) result in variations in the interior environment.

Exterior Environment

The exterior environment produces and influences the daylight that becomes the source of light for interior daylighting calculations. The constantly changing exterior environment is the key factor in the analysis of daylight. The three sources of light are the sun, sky, and ground.

Sun

The sun is the only known source of heat and light for the earth. It's diameter is more than 100 times that of the earth, and it is approximately 93 million miles from earth. The radiant energy or solar radiation received from the sun is transmitted in the form of short waves, including ultraviolet, visible, and infrared energy. Solar radiation is the most important source of heat, which is produced by the absorption of most of the radiant energy including light.

Only 50 to 60 percent of the radiant energy reaching the outer limits of the earth's atmosphere actually reaches the earth's surface. The percentage of energy reaching the earth varies with latitude, season, and cloud cover. Losses in radiant energy are due to selective scattering, diffuse reflection, and absorption in the atmosphere. Scattering and reflection account for approximately 70 percent of the loss; the remainder is due to absorption by water vapor.

The rotation of the earth around its polar axis causes daily changes, and the rotation of the earth around the sun causes seasonal changes (Figure 10-1). The two **Equinoxes** are the times when the sun's noon rays are directly vertical at the equator. The vernal equinox occurs in the spring (March 21 in the northern hemisphere, NH). The autumnal equinox occurs in the fall (September 23, NH).

The **solstices** are the times when the sun's noon rays are directly overhead at latitude 23½° north and south. The summer solstice in the northern hemisphere occurs on June 22 when the sun is directly overhead at the Tropic of Cancer. The winter solstice occurs on December 22 (NH) when the sun is directly overhead at the Tropic of Capricorn.

The position of the sun at any given instant of time is expressed in terms of two angles. **Altitude** or **solar altitude** is the vertical angle (or elevation) of the sun above the horizontal plane. **Azimuth** or **solar azimuth** is an angle measured in the horizontal plane. The horizontal angle is measured from due south to the vertical plane through the sun (see Figure 10-2).

Sky

Small obscuring particles (such as dust) and water vapor act to diffuse and scatter the radiant energy as it passes through the atmosphere. This results in

272

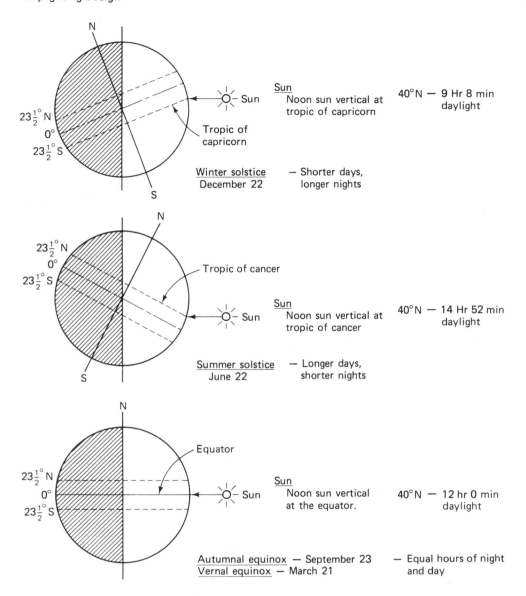

Figure 10-1. Seasonal Changes

what is referred to as **sky luminance.** The belt of maximum radiant energy moves back and forth across the equator as the seasons change and from east to west as the earth rotates. The luminance of the sky forms a vault or dome of nonuniform luminance. Sky luminance is evaluated in terms of **overcast** sky or **clear** sky luminance.

Overcast sky luminance varies as a function of location, time, density of

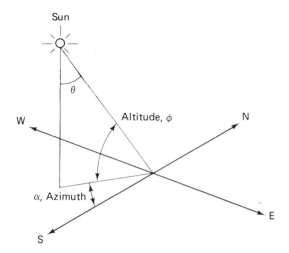

Figure 10-2.   Altitude and Azimuth

cloud cover, and uniformity of cloud cover. The luminance pattern is not uniform and will have a luminance about two and one half to three times greater at zenith (overhead) than on the horizon.

Clear sky luminance varies as a function of the location of the sun and the amount of atmospheric haze or dust. The luminance of a clear sky is greater near the horizon than at zenith except in the vicinity of the sun, where sky luminance increases.

Ground

Light reflected from the ground is a function of the reflective properties of the ground cover. The effectiveness of ground light is also dependent on the orientation (exposure to direct sun) of the window, obstructions between sun and ground and/or ground and window, and the height of the window above the ground.

Obstacles

Obstacles in the exterior environment may obscure contributions or increase reflected light from any one or more of the sources of light. Obstacles would include adjacent buildings or structures and landscape. High landscape may obscure contributions from the ground, sun, and/or sky. The blocking effect of high landscape, such as deciduous trees, may vary from season to season. Lower landscape or ground cover may obscure contributions from the ground and may also be seasonal, depending on the type of landscaping material.

Interior Environment

The daylight entering a space must be analyzed in terms of the quantity and quality of the light. Daylight may be sufficient in quantity to reduce the artificial lighting level needed, but result in false energy conservation if the quality of the

light is not analyzed. A poor quality of daylight may result in discomfort and a loss in visibility, which may cause a decrease in human performance and productivity. This, in turn, may cause increased use of the space, resulting in additional energy consumption.

Quantity

The footcandle is the unit associated with the quantity of daylight produced in a space (i.e., illumination). An illumination criterion is used in most methods of daylighting design because of the availability of instruments to measure footcandles and because of the relative ease of calculating illumination. Illumination levels may vary throughout the space at any given time depending on the number, location, and orientation of windows, the season, and exterior conditions. Because of limited advances in numerical methods for daylighting design, illumination criterion is the only technique in the literature available to the designer. The designer must be aware of the limitations of methods that deal only with the quantity of light. That is, illumination (footcandles) has little or no bearing (Chap. 7) on the ability of man to perform a visual task.

Quality

The quality of daylighting in a space affects its physical and psychological impact on the human occupants. To properly deal with the quality aspects of daylight, the human occupant must be included. Because of the complex nature of man, the analysis of quality is almost as variable as the stimulus (daylight). The physical effects of daylighting can be investigated in terms of the visual, thermal, and acoustical environment. The psychological effect of daylight is investigated in terms of man's subjective response to his environment. The emotional response to daylight by man is looked at in terms of very basic **like–dislike** and **need–don't need** modes of response.

PHYSICAL EFFECTS

Visual Environment. The quality of the visual environment in terms of daylighting is a very complex problem. One must deal with the response of the human visual system to daylight. The quality of visual stimulus is dependent on many factors. This text deals with the three most important factors that affect quality: (1) glare, (2) luminance ratios, and (3) color.

Glare can be defined as any excessively bright source of light within the visual field that creates discomfort and/or a loss in visibility. In layman's terms, discomfort is associated with pain, fatigue, strain, or increased tension. Loss of visibility is, as the term implies, a complete or partial loss in the ability to see the task or object of interest. As a glare source moves closer to the line of sight, both discomfort and loss of visibility increase as an exponential function. For example, as the bright lights of an oncoming automobile approach the driver's line of sight, the effect on visibility and comfort will become greater. Therefore, glare is a function of the source, location, intensity, surrounding luminance, and direction of view. If the occupant is involved in heads-up tasks (Chap. 7), direct glare will

be the phenomena of concern; occupants involved in heads-down tasks (Chap. 7) will be influenced by reflected glare or veiling reflections.

If the window is serving one of its primary psychological and emotional functions, that is, visual relief, it will be on the line of sight, which will compound the problems of direct glare. Direct glare from a source on the line of sight may cause discomfort depending on the differences in luminance between the window and the interior environment. The greater the luminance difference or luminance ratio, the greater the discomfort. Loss in visibility due to direct glare will be due to transient adaptation. Transient adaptation is a measure of the adaptation state of the visual system as the eye moves about the nonuniform environment. It takes time for the visual system to adapt to the variation in luminances within the luminous environment. Relative to daylight, the primary concern would be readaptation to the lower luminance of a visual task or object in the environment after subjecting the visual system to the potentially high luminance of the window. Adaptation from light to dark situations is referred to as **dark adaptation.** The luminance values in most interior environments would not be thought of as dark. However, relative to the potentially high levels of luminance of the window, the luminance difference could vary from a ratio of 1 to 1 to more than 10,000 to 1. Fortunately, this adaptation for most interior environments involves the cone receptors, which have a much quicker recovery time than the rods. The time it takes to readapt to the interior luminances is a function of the luminance difference between the window and the visual task, the length of exposure to the window luminance, the magnitude of the window luminance, and the size or visual angle subtended by the window. Readaptation time will increase with an increase in luminance difference, exposure, window luminance, and a decrease in window size.

The time it takes to readapt will result in a temporary loss in the ability to see. Most occupants will compensate for this loss in visibility by sacrificing speed or accuracy. Loss in speed and/or accuracy will result in a reduction in performance, and hence productivity. Although these losses may be small for a single glance at an excessively bright window, the accumulated losses over an 8-h day could be significant. When evaluating energy reduction by replacing artificial light by daylight, one must look at the potential increase in work time and energy consumed to compensate for the possible loss in productivity due to direct glare, transient adaptation, and readaptation.

Reflected glare can result from the reflection of the image of the window off a specular surface (Chap. 7). The problem of reflected glare involves source location and orientation of the task. Care in analyzing the location of the offending zone (Chap. 7) and proper orientation of tasks relative to windows will result in little or no problems with reflected glare.

Veiling reflections result in a loss in visibility due to a reduction in contrast (Chap. 7). Because of the strong unidirectional (unilateral placement) quality of most windows, the relationship between source and task location can be optimized. If the guidelines in Chap. 7 are followed, daylighting can result in increased

visibility or higher ESI values. A computer program could be used to study the visibility (ESI) of various tasks under different task–window orientations.

Luminance ratio is a ratio of the luminance of a task to the luminance of the area surrounding the task. In daylighting, the primary concern with regards to luminance ratios is between the luminance of the window and its immediate surrounding area of walls and/or mullions and frame. The visual system experiences increasing discomfort as the ratio between the window luminance and the surrounding area increases. The **IES Lighting Handbook**[1] (page 11-7) indicates that the maximum luminance ratio anywhere in the visual field should not exceed 40 to 1. The fifth edition[2] (page 11-3) states that ''for the best results the highest acceptable luminance of any significant surface in the visual field should not be greater than 5 times the luminance of the task.'' The 40-to-1 luminance ratio has been deleted from the latest edition. The author interprets the 40-to-1 ratio to be between the highest and lowest luminance level that can be seen from one location as an occupant casually looks around a space. It is interpreted as not being involved in looking back to a critical visual task. If the occupant is working on a critical visual task and looks up from the task, he should not be exposed to a luminance ratio greater than 5 to 1. If the ratio exceeds 5 to 1, a loss in visibility will occur as described previously.

Color is important in terms of the psychological response to color as well as color-rendering characteristics. Object color is seen by means of a phenomenon known as selective absorption. That is, object color is dependent on the pigment characteristics and the spectral distribution characteristics of the light source.

The spectral distribution characteristics of daylight[3] are made up of direct sunlight and the diffuse light from the sky. In general, the diffuse sky creates more short-wavelength energy than direct sunlight. The ratio of direct sunlight to diffuse skylight varies as a function of the solar altitude and aximuth, latitude, and atmospheric density and transmission. The thicker the atmosphere, the greater the effect on short-wavelength transmission through the atmosphere.

The spectral distribution characteristics of daylight may also be modified owing to transmission through different types of glass. Color-rendering effects of daylight can be a complex matter. Since the spectral distribution of daylight plus the spectral transmission of the window can vary over quite a range, one must be careful to examine colors under the appropriate conditions. Since most artificial light sources used for interior applications have vastly different spectral power distribution characteristics than sunlight, one must be extremely careful when selecting colors to be used under the combined artificial/daylight luminous environment.

Thermal Environment.   The thermal aspects of the interior environment as affected by the window are dealt with in detail later in this chapter. In general, the window affects the thermal environment in terms of heat transfer between the interior and exterior environments and in terms of heat exchange and distribution within the interior environment.

Acoustical Environment.   The effect of the window on the acoustical environ-ment will also be dealt with later in this chapter. In general, the window affects the acoustical environment in terms of sound transmission through openings around the glass or through the glass. Secondary noise production within the en-vironment can be generated owing to improper fittings that result in vibration or rattle. Also, smooth, hard surfaces such as glass can reflect sound, which may add to the ambient noise level.

PSYCHOLOGICAL AND EMOTIONAL EFFECTS.   Psychological and emotional aspects of daylighting have not been specifically investigated. Preliminary studies[4] have been made on the psychological effects of light on man in terms of artificial light. This type of research needs to be extended to include the effects of daylight on attitude, well-being, and motivation. The basic question is whether the space should have windows or be windowless. The problem is more than simply balanc-ing physical aspects of quality (visual, thermal, acoustic) against artificial lighting to produce energy conservation. One must be concerned with the way that daylight affects the user's spatial perception and behavior. If daylighting or the absence of daylighting has an adverse effect on attitude and motivation, it can have a detrimental effect on performance and productivity. Loss in performance and productivity may result in increased use of energy.

What are the effects of a windowless environment? In 1962, the Architec-tural Research Laboratory of the University of Michigan conducted a study of the effect of windowless classrooms[5] on the learning process. The test involved two elementary schools of the same type of construction. Mann Elementary School was used as the control situation with windows. Hoover Elementary was modified to be windowless for 1 year and a follow-up of 1 year with windows was included. Both elementary schools included kindergarten through third grade. Performance testing of the children in the windowless classroom indicated no ef-fect on the learning process when compared to students in the control school.

The study of windowless versus windowed environments needs to be ex-tended. In most cases where windows exist, occupants find that excessive direct sun and glare result in the use of a covering or shielding system, which turns the space into a windowless environment. A covered window seems to be less disturb-ing than a truly windowless room, since the occupant knows that the windows are there if he wants to see outside.

The psychological and emotional impact of the window or the absence of the window needs to be investigated to determine the effect of these factors on performance, productivity, attitude, behavior, and motivation.

Fenestration

The interface between the exterior environment and the interior environ-ment is the **fenestration.** The fenestration or openings are analyzed in terms of their application. The term **window** applies to all openings in the sidewalls, or sidelight; the term **toplight** applies to all systems utilizing an opening in the ceiling-roof plane, which provides overhead light. This section concentrates on the analysis of the physical characteristics of the most common interface material, glass. Some important characteristics of plastics are described as they apply to the interface for toplight.

Terminology and Units

1. Transparent glass (clear glass, vision glass): a material that transmits light without any apparent change in direction or color. Objects can be seen clearly through the material in either direction.

2. Translucent: a material that transmits light but diffuses the light as it passes through. Objects cannot be seen clearly through the material.

3. Opaque: a material that will not transmit light.

4. Reflective glass: a glass material that is coated on the outside with a transparent metallic oxide coating. During the daytime, when viewed from the inside the material appears transparent, and when viewed from the outside it appears to be opaque and acts as a mirror surface. The reverse is true at night.

5. Tinted glass: glass that contains additives that change the color and appearance, and reduce transmission.

6. U-coefficient of heat transmission (U-value): the number of British thermal units per hour (Btuh) that passes through 1 ft^2 of interface material when the temperature difference between interior air and exterior air is 1 °F for a steady rate of heat flow.

7. Conductivity, k: the number of Btuh that passes through 1 ft^2 of interface material 1 in. thick when the temperature difference ($\triangle T$) between interior and exterior is 1 °F ($\triangle T = 1$ °F) for a steady rate of heat flow.

8. Conductance, C: same as conductivity except for a specific thickness (x) or a thickness other than 1 in. ($C = k/x$).

9. Thermal resistance, R: the reciprocal of conductivity ($R = 1/k$); it is the number of hours required for 1 Btu to flow through interface material of k conductivity. For thicknesses other than 1 in. expressed as conductance, C, the thermal resistance would be x/k.

10. Air-space conductance, a: the rate of heat flow (Btuh) between the bounding surfaces through 1 ft^2 of area for $\triangle T = 1$ °F. It is affected by orientation and the emissivity, E, of the bounding surfaces.

11. Emissivity, E: the effective thermal absorption of the bounding surfaces of the air space.

12. Surface film conductance, h: conductivity dependent on the speed at which the air strikes the interface material. It is the rate of heat flow (Btu/h) through 1 ft^2 at $\triangle T = 1$ °F due to air motion across the surface:

outside: h_o = 4.0 Btu hr per ft² per °F for a 7.5-mph wind

inside: h_i = 1.46 Btu hr per ft² per °F

13. Infiltration: movement of air between interior and exterior environments due to cracks around window. The air flow is usually from cracks on the windward side to cracks on the leeward side.

14. Thermal pressure: the difference (ΔT) between interior (t_i) and exterior (t_o) temperature.

15. Shading coefficient (*SC*): a ratio of the solar heat gain of a sheet of glass to the solar heat gain of the reference glass. The reference glass is double-strength sheet glass with τ = 0.87, ρ = 0.08, and α = 0.05. Shading coefficient can be calculated as a ratio of the solar heat gain coefficient (*F*) of a sheet of glass to the *F* value for the reference glass. For single glass, $F = \tau + (U\alpha/h_o)$; for the reference glass, *F* = 0.87. Therefore,

$$SC = \frac{F \text{ of glass}}{0.87}$$

16. Greenhouse effect: trapping of solar radiation by the conversion of absorbed radiation by surfaces within a room into long-wavelength radiation. Most glass is opaque to long-wavelength radiation beyond 3 micrometers (μm).

17. Solar optical properties: transmittance (τ), reflectance (ρ), and absorptance (α). The total solar radiation striking a surface must equal the sum of the transmitted, reflected outward, and absorbed energy (Figure 10-3): $\tau + \rho + \alpha$ = 1.00. The magnitude of each of the three optical properties is a function of the thickness of the glass, physical properties of the glass, surface properties (coatings or film), and the angle of incidence, θ.

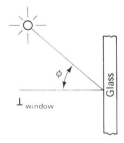
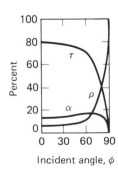

Figure 10-3. Variation in Optical Properties for Typical Quarter-Inch Plate Glass

Physical Characteristics of the Material

Both single glass and double or insulating glass may be clear, tinted, or reflective.

Clear glass is an optically transparent material while tinted glass contains additives that reduce the visual transmittance and shading coefficient and modify the color appearance. Reflective glass consists of a transparent metallic oxide coating applied to the surface of the clear or tinted glass. For single sheet glass the reflective coating is normally placed on the outside surface. For double or insulating glass, the reflective coating (see Figure 10-4) can be placed on the outside surface of the outdoor glass, or on the air-space surface of the outdoor glass, or on the air-space surface of the indoor glass. The outdoor glass and indoor glass can both be clear or the outdoor glass can be tinted while the indoor glass is clear.

Glass has a specific gravity of 2.40 to 2.60 and weighs approximately 156 to 158 lb/ft³. This is heavy compared to most building material. For example, concrete has a weight of 144 lb/ft³, and common brick weighs 120 lb/ft³. A more meaningful way to express the weight of glass is in terms of its weight per square foot of surface area. For example, a ¼-in.-thick glass will weigh between 3.25 and 3.29 lb/ft².

The strength of glass is a key consideration in its selection. The strength attributes are expressed in terms of the thickness of glass required. The thickness is a function of the external pressure loading (wind, sonic, etc.), the surface area of the glass, strength properties of the glass, surface characteristics of the glass, and the support conditions.

Environmental Impact of the Window

THERMAL CHARACTERISTICS.   The thermal characteristics of glass with or without shading devices are an important property that must be considered. The heat gain and/or heat loss properties of glass will have an influence on the initial and

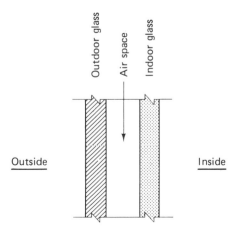

Figure 10-4.   Double-Glass Terminology

operating cost of the mechanical conditioning system, which in turn will influence the level of human comfort and performance. Increases in heat gain (or loss) will occur from an increase in the shading coefficient, U-value, and air-to-air temperature difference.

The U-value represents the reciprocal of the sum of the thermal resistance values of each material, including air spaces and surface films. Table 10-1 gives an indication of the change in U-value for various conditions.

TABLE 10-1

Typical U-Values[6]

	Summer[a,c]	*Winter*[b,c]
Single glass	1.06	1.16
Double glass	0.61	0.65
Triple glass	0.45	0.47

[a]Outside air film, summer at 7½ mph; $h_o = C = 4.0$.
[b]Outside air film, winter at 15 mph; $h_o = C = 6.0$.
[c]Inside air film, still air, $h_i = C = 1.46$.

One manufacturer lists the U-value for a single sheet of glass as 1.0 for winter conditions (15 mph), which results in $R_{glass} = 0.15$. If $U = 1.0$, then

$$R_{tot} = \frac{1}{1.0} = 1.0$$

If in winter $h_o = C = 6.0$, then

$$R_{h_o} = \frac{1}{6.0} = 0.17$$

If for the interior, $h_i = C = 1.46$, then

$$R_{h_i} = \frac{1}{1.46} = 0.68$$

Therefore, $R_{surface\ films}$ $= 0.85$.

$$R_{glass} = R_{tot} - R_{surface\ films}$$
$$= 1.0 - 0.85 = 0.15$$

In Table 10-1, the resistance of glass is

$$R_{glass} = 0.01 \quad (U = 1.16)$$

An example in the **ASHRAE Handbook of Fundamentals**[7] gives the resistance of glass as

$$R_{glass} = 0.035 \quad (U = 1.13)$$

This gives a variation in U-value from 1.0 to 1.16, with an apparent variation in R

from 0.01 to 0.15, which appears to be an inconsistency in the literature. The lower the *U*-value, the lower the heat gain or heat loss.

SOUND CHARACTERISTICS.   The sound transmission loss for various types of glass varies from approximately 24 to 40 decibels (dB) for frequency range of 125 to 4000 hertz (Hz). The transmission of noise from the exterior environment into the interior environment is dependent on the air tightness of the building. Street or traffic noise transmitted through and around glass can be annoying and distracting. In addition to exterior noise transmission, the window and frame combination may create noise or rattle owing to vibration of improperly fitted materials.

MAINTENANCE CHARACTERISTICS.   Maintenance of the glass is important to maintain the level of light transmission and desirable visibility. The length of time between cleaning periods is a function of the internal and external environmental conditions. Maintenance of the interior surface of the glass will usually be more convenient, which should result in a more repetitive cleaning cycle. For multistory, fixed window systems, exterior cleaning is more difficult and costly resulting in longer cleaning cycles. With longer cleaning cycles for exterior glass surfaces, loss of light transmission can vary substantially depending on the characteristics of the exterior environment. The percent of transmission loss for one study (reference 1, Figure 7-31) indicated a variation in light loss of from 73 percent for a "typical clean location" to 55 percent for a "typical dirty location" after 6 months.

For example,

$$\tau_{glass} = 85\%$$
$$\tau_{glass} = 0.85 \times 0.73 = 0.62 = 62\%, \text{ clean location}$$
$$= 0.85 \times 0.55 = 0.47 = 47\%, \text{ dirty location}$$

It is obvious that this loss in transmission can have quite an effect on the quantity of illumination reaching the interior space. Dirt or film deposits on the exterior of the glass surface will also affect the quality of the light and the mode of heat transmission. The quality of the light will probably be more diffuse in nature, resulting in more nondirectional scatter of the light entering the space. Heat transmission will be affected due to the change in the surface air film characteristics, which will affect the h_o value.

Abrasive cleaners should be avoided when cleaning glass to prevent scratches. Fingerprints, grease, dirt, scum, and glazing materials can be removed with a wash, rinse, and dry technique recommended by the manufacturer. A mild soap, detergent, or slightly acidic cleaning solution can be used to wash most glass surfaces. For some types of glass and surface grime, mild commercial solvent can be used. Glass with reflective coatings can be damaged or scratched if not cleaned properly. Alkaline or fluorine material that is associated with concrete, masonry, or decorative crushed rock can stain or etch the glass. Oxide deposits from weathering steel or rust can also stain or etch glass.

Maintenance of toplight or skylight systems can be more critical in terms of accessibility and dirt accumulation. The horizontal orientation of a toplight system will accumulate dirt more rapidly than a vertical surface. Cleaning of the toplight from the outside may be easier than cleaning the outside of vertical windows; however, accessibility to interior toplight surfaces may be more difficult, and require ladders or scaffolding to be brought into the space, which may discourage cleaning. Deposits of dirt and surface film materials such as smoke present in the environment will adversely affect the efficiency of transmission and the distribution characteristics. Static charges, which are common with plastic surfaces found in toplight systems, may compound the maintenance problem.

Control Elements

Control elements are introduced to prevent the direct transmission of the sun, to reduce glare, and to reduce heat gain and heat loss. The control elements can be internal or external. They can be an integral part of the window or a separate element from the window.

Direct transmission of radiant energy from the sun should be prevented to minimize luminance ratios within the interior environment, as well as to prevent color fading. Both interior and exterior shielding elements can be used. Exterior shielding elements would include overhangs, louvers, building elements or projections, and landscaping. Interior shielding usually consists of shades, blinds (vertical and horizontal), and draperies. Interior shielding has the disadvantage of allowing the penetration of the radiant energy into the space, which will be converted to heat and result in a heat gain due to the greenhouse effect. Interior shielding is used more often than exterior shielding because of its accessibility and the ease of maintenance. To be effective as a shielding device, the material must be opaque or of very low transmittance.

Glare control can be achieved by reducing the apparent luminance of the window surface or by moving the glare source out of the direct field of view. Reductions in luminance or luminous intensity entering the interior environment can be achieved by using tinted or low-transmission glass or by using low-transmission transluscent shades, blinds, or draperies. Adequate control of glare may result in excessive loss in luminous energy entering the space, causing a reduction in the effectiveness of the window as a source of light. Glare control may also adversely affect the ability of one to look out of the window and may produce unacceptable color distortion because of selective transmission. Direct glare can be reduced by moving the daylight source overhead, as in toplighting. However, direct solar transmission must also be controlled when using toplighting. Also, because of the strong directional quality of the light from toplight, reflected glare and veiling reflections may be problems.

Heat gain can be reduced by using tinted and/or reflective thermal glass. Double and triple glass systems utilizing reflective coatings and tinted glass have been introduced to reduce heat loss and heat gain. Visibility through the glass and color perception should be considered when selecting thermal glass. The consequences of reflecting radiant energy from the building to adjacent property owners must be investigated.

A dual-mode shading device[8,9] has been investigated by Silverstein. The system consists of a reversible roller shade that has a dark-colored absorbing side while the other side is a solar reflector. The system provides for the control of direct sunlight, the control of glare, and the control of heat gain and heat loss. The dark side is turned toward the sun during winter months to act as a solar collector that, when combined with the proper natural ventilation, allows the heat to be circulated into the space. The air space created by the shade adds to the insulating capabilities of the system during winter evenings. The reflective side of the shade can be used to reject solar heat gain during the summer. To increase its efficiency during the summer, an interior–exterior air circulation system should be provided.

Rosenfeld[10] has suggested a modification to the Silverstein system. He has suggested using venetian blinds (see Figure 10-5) constructed of low-transmission gray plastic with a metallic reflective coating on one side. The primary advantage of the venetian blind is that the blinds can be opened slightly to allow for more natural ventilation during the summer, and on cloudy days they can be partially opened to admit more daylight.

A beam-daylighting system[11] has been investigated to increase the effective utilization of daylighting by increasing daylighting penetration in the space. The system utilizes a venetian blind with a metallic reflective coating on one side. The

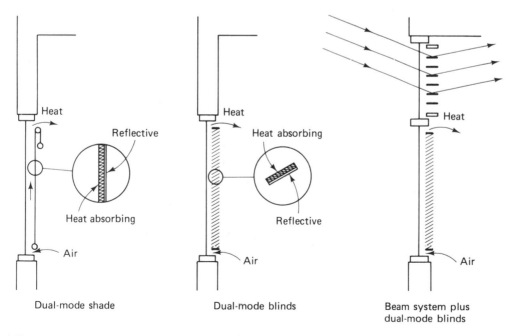

Figure 10-5. Daylight and Heat Control Devices

silvered beam blind is mounted above eye level and utilizes direct solar radiation incident on the top 1 to 2 ft of the window (see Figure 10-5). The beam blind is separate from the lower window shading system to allow for independent action. The purpose is to reflect the solar radiation from the beam blind to the diffuse white ceiling plane. This indirect lighting system provides greater daylight penetration with more uniform diffusion. The sun orientation, blind tilt, and reflective properties of the ceiling are critical to the efficient utilization of the daylight. Since this system allows for the penetration of direct solar radiation (redirected to the ceiling) into the space, it will result in direct solar heat gain to the interior environment. If the system is not performing at its optimum, sufficient daylighting may not be present to allow for a sufficient quantity of artificial lights to be switched off to offset the heat gain produced by the direct solar radiation.

DESIGN OF DAYLIGHTING

An analysis of the important aspects of the exterior environment, interior environment, and fenestration is essential to the beginning of the design process. Each opening placed in the exterior wall of a building should be evaluated in terms of its total impact on the interior environment and the human occupants. The decision to place fenestration in a building should not be based on arbitrary,

Advantages	
Interior/exterior visual communication	Indoor plant growth
Design enhancement: rhythm, relief, drama, pattern, etc.	Fire escape
	Security: easier to observe interior when
Ventilation[a]	entering the space
Opens the space, feeling of enlargement	Awareness of exterior environmental conditions
Heat gain in winter	Visual and psychological relief: varying
Potentially good ESI with proper task placement (see Chap. 7)	source of light, changing color, changing mood

Disadvantages	
Unpleasant view	Condensation
Glare source	Undependable, constantly varying light
Potential color fading and ultraviolet damage	source
	Shadowing from strong direction source
Heat loss in winter	Sound transmission[a]
Heat gain in summer	Security: easily penetrated
Bodily discomfort: radiation loss due to surface temperature	Air pollution, if operable for ventilation[a]
	Limits circulation
Increased maintenance: cleaning and breakage	
Higher cost than conventional wall materials	

[a]Ventilation will increase sound transmission and the amount of air pollution entering the space.

superficial reasoning. The table on page 286 lists the advantages and disadvantages of placing fenestration in the exterior walls.

The list is not complete. Each designer may not agree with all factors and should compile a personal list. However, this type of listing procedure accompanied by a study of the analysis section should give the designer sufficient information to make a decision for or against placing fenestration in a room. This decision may vary from room to room, depending on the function of the room, its orientation, and its occupants. This statement implies a potentially drastic change in the outward appearance of buildings to more appropriately express the inward function, while weighting the impact on energy conservation.

Once the decision has been made to provide fenestration in a space, the size, proportion, and placement of the opening must be decided by the designer. The size should be based on the thermal impact of the opening and the daylighting potential. The proportions and placement should be based on the quality and quantity of daylight entering the space. The following factors should be considered in window design:

1. Avoid the direct penetration of sunlight.
2. For low buildings of one or two stories, use light-reflecting ground cover to improve light penetration into the space.
3. Provide openings in more than one wall if possible to improve penetration, uniformity, and reduce harsh shadows.
4. Use high, clear glazing; the head of the window is most effective in admitting light to the inner portion of a space.
5. Two rules of thumb:
    a. Optimum daylighting penetration and uniformity will be achieved if the window height is at least one half the room depth.
    b. Glass area should be approximately equal to 25 percent of the floor area to optimize the uniformity of illumination in a space.
6. The reflectance properties of external shielding devices can be critical to daylight penetration.

The following factors should be considered in toplight or skylight design:

1. Avoid the direct penetration of sunlight.
2. For monitor roof, clerestory, and sawtooth configurations, the use of light-reflecting roof materials will increase the quantity of light entering the space.
3. Rules of thumb:
    a. Maximum spacing between skylights is 1.5 times the ceiling height.
    b. Skylights should occupy approximately 5 percent of the ceiling area; this percentage is recommended to minimize ceiling clutter when skylights (5 percent) are combined with ceiling-mounted luminaires (10 to

20 percent coverage) to allow for a 75 to 85 percent ceiling surface area, which is important to the interreflection of energy.

4. Openings through a flat roof plane create flashing problems that are difficult to solve because of extreme temperature variations, causing differential expansion and contraction.

5. Heat buildup due to natural heat convection currents in a room may result in excessive heat loss.

6. Use moisture seals for condensation prevention and drainage.

Assuming suitable sun control, a southern exposure (continental United States) is preferred to optimize the daylight contribution into a space. Sufficient daylighting must be provided to replace artificial lighting if energy conservation is to be realized. Northern exposure may not provide a sufficient quantity of illumination to allow for a significant reduction in artificial lighting. In general, east–west orientations should be avoided since they present the most difficult problems in daylight control and create complicated, extreme heating and cooling problems.

Longhand methods of calculating illumination levels in a room are given in the example problems at the end of this chapter. A computerized design technique has been developed. Although more complex, the engineer is urged to become familiar with the program. Because of the speed of the computer, a more complete and accurate profile of the daylight contribution can be obtained.

The longhand design procedures involve two steps: (1) determining the quantity of illumination coming to the window surface, and (2) using that quantity to determine the daylight contribution to the interior part of the space. To determine the quantity of illumination reaching the window, one of the following design conditions must be assumed.

1. Clear day (surface exposed to the sun): the design is based on light reaching the window from three sources.

   a. Clear sky

   b. Direct sun

   c. Ground contribution

2. Clear day (nonexposed surface)

   a. Clear sky

   b. Ground contribution

3. Overcast day

   a. Overcast sky

   b. Ground contribution

Two procedures are given for determining the quantity of illumination coming to the window surface and its effect on interior illumination. The first procedure is based on data supplied by the Illuminating Engineering Society[2]; the second procedure is based on data presented by Kinzey and Sharp[12].

IES Daylight Procedure

Clear skylight illumination values given in the data section at the end of this chapter are based on an equivalent clear-sky luminance. The single illumination value represents an average of the luminance patterns of the sky for a specific time, date, latitude, and compass direction. The clear-sky luminance is assumed to be uniformly distributed across the sky. This assumption of a single equivalent clear-sky luminance value to describe the illumination contribution to the window surface is questionable. The direct-sun contribution to the surface of a window is given in the data section also. The illumination values in the table are based on average values. The illumination is on a surface normal to a line from the sun to the window surface. To find the actual level of illumination on the vertical surface of the window, the value must be multiplied by the correct trigonometric functions (see Figure 10-6).

$$E_{\text{window}} = E_{\text{sun}} \times \cos \phi \times \cos \psi$$

where ϕ = altitude

ψ = angle between the perpendicular to the window and the vertical plane through the sun

$\psi = \alpha_1 - \alpha_2$

α_1 = solar azimuth

α_2 = azimuth angle to a perpendicular to the window

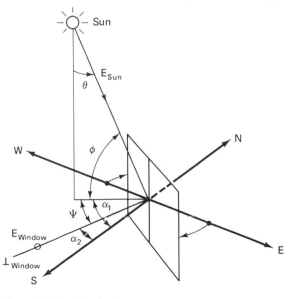

Figure 10-6. Illumination on the Plane of the Window

The illumination produced on a window surface from an overcast sky can be found in the data section. Average equivalent sky luminance values for an overcast day are based on the average distribution of luminance across the sky.

Ground-light contribution is dependent on the reflectance of the ground cover, the distance between the window and ground, and obstructions. For a ground of infinite extent with uniform luminance, the illumination contribution from the ground can be assumed to be equal to one half the ground luminance. The ground luminance is calculated by multiplying the horizontal illumination on the ground by the reflectance value of the ground-cover material. For limited ground areas, a more exact method is necessary.

The solar altitude and solar azimuth values must be determined to begin the design process. These values are based on latitude, season, and time of day. Altitude is the vertical angle from the horizon to the sun; azimuth is the horizontal angle measured from south to the plane through the sun (Figure 10-2).

Once the contribution of illumination to the window surface has been calculated, two longhand methods are available for determining the illumination contribution to the space. The first method involves the calculation of illumination at points in the room. An example of this method will be found in the example problems. The point-by-point procedure makes two assumptions that affect the accuracy of the calculation: (1) the interreflected component is ignored, and (2) the window is a uniform diffuse emitter. The second method is a lumen method that calculates illumination values at three points, defined as the maximum, midway, and minimum. An example of this method can be found in reference 2. The second procedure includes both the direct and interreflected components of illumination. It also assumes that the illumination coming to the window is uniformly distributed across the surface of the window.

Alternative Procedure[12]

The second procedure consists of determining the illumination on the surface of the window, which is modified by coefficients in a lumen formula (see Figure 10-7). The method uses a window ratio to find minimum, maximum, and average illumination coefficients. The general formula for calculating the minimum, maximum, and average illumination in a room is

$$E = \frac{E_v \times A_w \times K \times T_g \times \text{LLF}}{A_f}$$

where E_v = illumination on the window
A_w = area of the window (minus frames and mullions)
K = illumination coefficient
T_g = transmission of the glass
LLF = light-loss factor
A_f = area of the floor

The data apply only for contributions from a window along one entire wall

General lighting level:

$$E_G = \frac{1}{3} \frac{(150 \times 450) + (70 \times 225)}{450 + 225} = 41.0 \text{ fc}$$

$$\text{NTA} = A_G - \Sigma A_T = 1350 - 675 = 675 \text{ ft}^2$$

$$A_G = \text{NTA} = \Sigma A_T = 675 \text{ ft}^2$$

$$\text{watts } (A_G) = \frac{41.0 \times 675}{55 \times 0.65 \times 0.70} = 1109 \text{ W}$$

$$\text{NCA} = \text{NTA} - A_g = 675 - 675 = 0$$

$$\text{total watts} = 2697 + 630 + 1109 = 4436 \text{ W}$$

3. Private office: 102

$$L = 30 \text{ ft}, \qquad W = 21 \text{ ft}, \qquad A_G = 630 \text{ ft}^2$$

$$h_{rc} = 5.5, \qquad \text{RCR} = \frac{5(5.5)(30 + 21)}{30 \times 21} = 2.23, \qquad \text{CU} = 0.60$$

T1. Two work locations at 70 fc.

$$A1 = 2 \times 50 = 100 \text{ ft}^2$$

$$\text{watts } (A1) = \frac{70 \times 100}{55 \times 0.60 \times 0.70} = 303 \text{ W}$$

Since, $\qquad \Sigma A_T = 100 < \frac{1}{2} A_G = 315,$

Therefore, $\qquad AA$ is 50 ft^2.

General lighting level:

$$E_g = \frac{1}{3}(70) = 23.3 \text{ fc}$$

$$\text{NTA} = E_G - \Sigma A_T = 630 - 100 = 530 \text{ ft}^2$$

Therefore, $\quad A_g = \Sigma A_T = 100 \text{ ft}^2$

$$\text{watts } (A_G) = \frac{23.3 \times 100}{55 \times 0.60 \times 0.70} = 101 \text{ W}$$

$$\text{NCA} = \text{NTA} - A_G = 530 - 100 = 430 \text{ ft}^2$$

$$E_{NC} = \frac{1}{3} E_G = 7.8 \text{ fc},$$

Therefore,

$$\text{ENC} = 10 \text{ fc}$$

$$\text{watts } (A_{NC}) = \frac{10 \times 430}{55 \times 0.60 \times 0.70} = 186 \text{ W}$$

$$\text{total watts} = 303 + 101 + 186 = 590 \text{ W}$$

II. Computerized Power Calculation

The following is a sample terminal input/output for a program developed by the author and a sample of worksheet for data input.

```
PROJECT NAME
? XYZ CORP.
PROJECT NUMBER(I5)
? 75000
```

Data Input
(from a computer terminal)

```
  ROOM NUMBER (I3)
? 100
WIDTH
? 40
LENGTH
? 66
HEIGHT OF ROOM CAVITY
? 5.5
EFFICACY
? 55
  LUMINAIRE SEQ(I5) NO.-RIGHT ADJ.
? DB117
NUMBER OF TASKS(I1)
? 3
TASK 1 — E
? 100
TASK 1 — WORK     LOCATIONS
? 5.
TASK 2 — E
? 70.
TASK 2 — WORK     LOCATIONS
? 16.
TASK 3 — E
? 30.
TASK 3 — WORK     LOCATIONS
? 4.
```

during the on period to compensate for variations in the daylighting contribution (compensating system) must be completely automatic. The compensating system could involve an on–off mode or a continuously varying mode (dimming). In either mode, the compensating system should have a delay circuit that would prevent response to momentary daylight reductions caused by cloud movement. Of the two compensating systems, the continuously varying mode would be preferred. The cycling effect of the on–off mode would be less expensive and better suited for retrofit applications, but it could also be quite distracting and annoying. The user system, which is the on–off control of the compensating system, should be controlled with a time clock or other device that would automatically turn off the compensating system at the end of the day. A short-period (maximum 3 h) manual override would have to be available for those who must work overtime or come in at night. In larger rooms, multiple dimmers and photocell sensors may be required to control separate rows of luminaires to assure a more uniform lighting level throughout the space.

Because of the current state of the art in lighting control systems, the initial cost of such a system would be quite high. However, as technology advances in this field, the competitive market should bring the cost down to a more reasonable level. Also, the cost of this type of system must be evaluated in terms of its payback period, owing to the reduction in operating cost, and in terms of its overall comfort to the user.

references

1. **IES Lighting Handbook,** 4th ed., Illuminating Engineering Society, New York, 1966.

2. Kaufman, John, editor, **IES Lighting Handbook,** 5th ed., Illuminating Engineering Society, New York, 1972.

3. Kimball, H. H., "Records of Total Solar Radiation Intensity and Their Relation to Daylight Intensity," **Transactions of IES,** May 1925, pp. 477–497.

4. Flynn, J. E., "The Psychology of Light: Series 1," **Electrical Consultant,** Dec. 1972–July 1973.

5. "The Effect of Windowless Classrooms on Elementary School Children," Architectural Research Laboratory, Department of Architecture, University of Michigan, 1965.

6. McGuinnes, W., and B. Stein, **Mechanical and Electrical Systems for Buildings,** 5th ed., John Wiley & Sons, Inc., New York, 1971.

7. **ASHRAE Handbook of Fundamentals,** American Society of Heating, Refrigerating, and Air Conditioning Engineers, Inc., New York, 1972.

8. Berman, S. M., and S. D. Silverstein, "Energy Conservation and Window Systems," AIP Conference Proceedings No. 25, American Institute of Physics, New York, 1975.

9. Silverstein, S. D., "Efficient Energy Utilization in Buildings: The Architectural Window," Proceedings 10th Intersociety Energy Conversion Engineering Conference, 1973, p. 685.

10. Rosenfeld, A. H., "Some Comments on Dual Solar-Control Venetian Blinds," Lawrence Berkeley Laboratory, Department of Physics, University of California, Berkeley, Calif.

11. Rosenfeld, A. H., and S. E. Selkowitz, "Beam and Diffuse Daylighting, and Peak Power," Proceedings: The Basis for Effective Management of Lighting Energy Symposium, Federal Energy Administration, Washington, D.C., Oct. 29, 30, 1975.

12. Kinzey, B. Y., and H. M. Sharp, **Environmental Technologies in Architecture**, Prentice-Hall, Inc., Englewood Cliffs, N.J., 1964.

data section

TABLE 10-2

Solar Altitude and Aximuth (Courtesy of the Illuminating Engineering Society of North America)

			Solar Time*						
	Date	AM:	6	7	8	9	10	11	Noon
		PM:	6	5	4	3	2	1	
30°N ALTITUDE	June 21		12	24	37	50	63	75	83
	Mar.–Sept. 21		–	13	26	38	49	57	60
	Dec. 21		–	–	12	21	29	35	37
AZIMUTH	June 21		111	104	99	92	84	67	0
	Mar.-Sept. 21		90	83	74	64	49	28	0
	Dec. 21		–	60	54	44	32	17	0
34°N ALTITUDE	June 21		13	25	37	50	62	74	79
	Mar.-Sept. 21		–	12	25	36	46	53	56
	Dec. 21		–	–	9	18	26	31	33
AZIMUTH	June 21		110	103	95	90	78	58	0
	Mar.-Sept. 21		90	82	72	61	46	26	0
	Dec. 21		–	–	54	43	30	16	0
38°N ALTITUDE	June 21		14	26	37	49	61	71	75
	Mar.-Sept. 21		–	12	23	34	43	50	52
	Dec. 21		–	–	7	16	23	27	28
AZIMUTH	June 21		109	101	90	83	70	46	0
	Mar.-Sept. 21		90	81	71	58	43	24	0
	Dec. 21		–	–	54	43	30	16	0
42°N ALTITUDE	June 21		16	26	38	49	60	68	71
	Mar.-Sept. 21		–	11	22	32	40	46	48
	Dec. 21		–	–	4	13	19	23	25
AZIMUTH	June 21		108	99	89	78	63	39	0
	Mar.-Sept. 21		90	80	69	56	41	22	0
	Dec. 21		–	–	53	42	29	15	0
46°N ALTITUDE	June 21		17	27	37	48	57	65	67
	Mar.-Sept. 21		–	10	20	30	37	42	44
	Dec. 21		–	–	2	10	15	20	21
AZIMUTH	June 21		107	97	88	74	58	34	0
	Mar.-Sept. 21		90	79	67	54	39	21	0
	Dec. 21		–	–	52	41	28	14	0
48°N ALTITUDE	June 21		17	27	37	47	56	63	65
	Mar.-Sept. 21		–	10	20	29	36	40	42
	Dec. 21		–	–	1	8	14	17	19
AZIMUTH	June 21		106	95	85	72	55	31	0
	Mar.-Sept. 21		90	79	67	53	38	20	0
	Dec. 21		–	–	52	41	28	14	0

*Time measured by the daily motion of the sun. Noon is taken as the instant in which the center of the sun passes the observer's meridian.

TABLE 10-3

Average Solar Illumination as a Function of Altitude

Latitude	Plane	Illumination (footcandles)								
		December 21			March 21, September 21			June 21		
		8 A.M. 4 P.M.	10 A.M. 2 P.M.	Noon	8 A.M. 4 P.M.	10 A.M. 2 P.M.	Noon	8 A.M. 8 P.M.	10 A.M. 2 P.M.	Noon
30°N	Perp.[a]	4200	7000	7700	6400	8300	8600	7700	8600	8900
	Horiz.	700	3400	4400	2600	5900	7000	4400	7200	8500
34°N	Perp.	3100	6500	7100	6300	8100	8400	7600	8600	8900
	Horiz.	400	2700	3700	2400	5600	6700	4700	7100	8400
38°N	Perp.	2500	6000	6900	6100	8000	8300	7600	8500	8900
	Horiz.	100	2000	3000	2100	5400	6200	4400	7000	8300
42°N	Perp.	2000	5500	6400	6000	7800	8200	7600	8400	8800
	Horiz.	100	1600	2700	2000	4800	5800	4700	6800	7900
46°N	Perp.	500	4500	5800	5800	7600	8100	7600	8100	8800
	Horiz.	—	1000	1800	1800	4400	5500	4400	6700	7400

[a]Perpendicular to sun's rays.

Courtesy of the Illuminating Engineering Society of North America.

TABLE 10-4

Transmittance Data of Glass and Plastic Materials

Material	Approximate transmittance (%)
Polished plate/float glass	80–90
Sheet glass	85–91
Heat-absorbing plate glass	70–80
Heat-absorbing sheet glass	70–85
Tinted polished plate	40–50
Reflective glass	23–30
Figure glass	70–90
Corrugated glass	80–85
Glass block	60–80
Clear plastic sheet	80–92
Tinted plastic sheet	42–90
Colorless patterned plastic	80–90
White translucent plastic	10–80
Glass fiber reinforced plastic	5–80
Double glazed, two lights clear glass	77
Tinted plus clear	37–45
Reflective glass[a]	5–25

[a]Includes single glass, double-glazed units, and laminated assemblies. Consult manufacturer's material for specific values.

Courtesy of the Illuminating Engineering Society of North America.

TABLE 10-5

Typical Light Loss Factors for Daylighting Design

| Locations | Light loss factor glazing position | | |
	Vertical	Sloped	Horizontal
Clean areas	0.9	0.8	0.7
Industrial areas	0.8	0.7	0.6
Very dirty areas	0.7	0.6	0.5

Courtesy of the Illuminating Engineering Society of
North America.

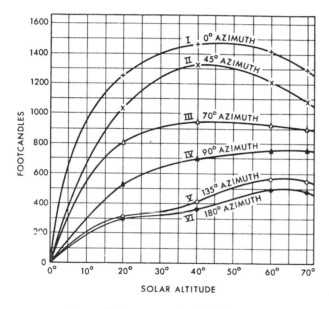

Curves of clear summer sky light illumination on vertical
surfaces:

Curve I. Vertical surface facing 0° in azimuth from sun.

Curve II. Vertical surface facing 45° in azimuth from sun.

Curve III. Vertical surface facing 70° in azimuth from sun.

Curve IV. Vertical surface facing 90° in azimuth from sun.

Curve V. Vertical surface facing 135° in azimuth from sun.

Curve VI. Vertical surface facing 180° in azimuth from sun.

Figure 10-9. Summer Sky Illumination (Courtesy of the Illuminating Engineering Society of North America)

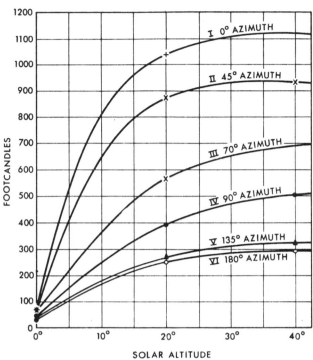

Curves of clear winter sky light illumination on vertical surfaces:

Curve I. Vertical surface facing 0° in azimuth from sun.

Curve II. Vertical surface facing 45° in azimuth from sun.

Curve III. Vertical surface facing 70° in azimuth from sun.

Curve IV. Vertical surface facing 90° in azimuth from sun.

Curve V. Vertical surface facing 135° in azimuth from sun.

Curve VI. Vertical surface facing 180° in azimuth from sun.

Figure 10-10. Winter Sky Illumination (Courtesy IES)

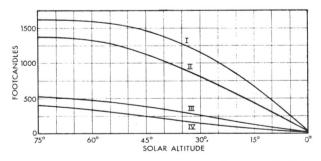

Illumination with a cloudy sky. Curve I, average for Plains states. Curve II, average for Atlantic coast states. Curve III, minimum for Plains states. Curve IV, minimum for Atlantic coast states.

Figure 10-11. Cloudy Sky Illumination (Courtesy IES)

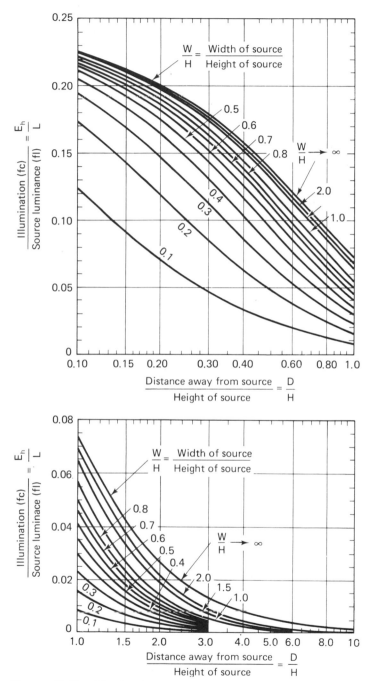

Figure 10-12. Illumination from a Rectangular Uniform Source Perpendicular to the Illumination Plane (Courtesy of Illuminating Engineering Society of North America)

TABLE 10-6

Window Ratio for Light from Sky or Sun

Hcw/RW [b]	RL/RW [a]			
	1	2	3	4
0.25	2.6	2.9	3.2	3.55
0.5	1.5	1.7	1.85	2.0
0.75	0.96	1.15	1.20	1.35
1.0	0.68	0.82	0.90	0.98
1.25		0.62	0.70	0.75
1.50				0.60

[a] RL = room length; RW = room width perpendicular to window.

[b] Hcw = center of window above floor.

(Reprinted with permission from Kinzey and Sharp, *Environmental Technologies in Architecture*, Englewood Cliffs, N.J.: Prentice-Hall, Inc., 1964)

TABLE 10-7

Window Ratio for Light from Ground

Hc/RW [b]	RL/RW [a]			
	1	2	3	4
0.25	3.3	3.9	4.5	5.0
0.50	1.8	2.15	2.5	2.75
0.75	1.3	1.50	1.7	1.85
1.0	0.98	1.20	1.3	1.43
1.25		0.95	1.12	1.18
1.50		0.75	0.88	0.97
1.75		0.65	0.75	0.82
2.0			0.68	0.78

[a] RL = room length; RW = room width perpendicular to window.

[b] Hc = ceiling height.

(Reprinted with permission from Kinzey and Sharp, *Environmental Technologies in Architecture*, Englewood Cliffs, N.J.: Prentice-Hall, Inc., 1964)

TABLE 10-8

Illumination Coefficients for Light from Clear Sky

Window Ratio	Illumination coefficient					
	Average (K_{ave})		Minimum (K_{min})		Maximum (K_{max})	
	70%	30%	70%	30%	70%	30%
0.50	0.65	0.41	0.57	0.32	0.68	0.45
1.00	0.69	0.50	0.55	0.31	0.80	0.60
1.50	0.73	0.55	0.52	0.30	0.96	0.82
2.00	0.78	0.60	0.50	0.28	1.21	1.10
2.50	0.82	0.65	0.48	0.25	1.53	1.38
3.00	0.86	0.70	0.46	0.22	1.89	1.75
3.50	0.90	0.75	0.43	0.20	2.27	2.15
4.00	0.93	0.79	0.41	0.19	2.64	2.53
4.50	0.96	0.81	0.39	0.18	3.05	2.93
5.00	1.00	0.82	0.37	0.17	3.45	3.30

Note: Ceiling reflectance, 85%; floor reflectance, 30%; wall reflectances, 70% and 30% as listed.

(Reprinted with permission from Kinzey and Sharp, *Environmental Technologies in Architecture,* Englewood Cliffs, N.J.: Prentice-Hall, Inc., 1964)

TABLE 10-9

Illumination Coefficients for Light from Overcast Sky

Window ratio	Illumination coefficient					
	Average (K_{ave})		Minimum (K_{min})		Maximum (K_{max})	
	70%	30%	70%	30%	70%	30%
0.50	0.65	0.53	0.53	0.40	0.72	0.60
1.00	0.72	0.59	0.48	0.35	0.90	0.79
1.50	0.75	0.64	0.42	0.30	1.16	1.05
2.00	0.81	0.69	0.37	0.25	1.55	1.42
2.50	0.85	0.73	0.32	0.20	1.98	1.84
3.00	0.89	0.77	0.29	0.16	2.48	2.32
3.50	0.91	0.78	0.24	0.12	2.96	2.78
4.00	0.91	0.78	0.19	0.09	3.48	3.20
4.50	0.91	0.77	0.15	0.06		
5.00	0.90	0.77	0.13	0.05		

Note: Ceiling reflectance, 85%; floor reflectance, 30%; wall reflectances, 70% and 30% as listed.

(Reprinted with permission from Kinzey and Sharp, *Environmental Technologies in Architecture,* Englewood Cliffs, N.J.: Prentice-Hall, Inc., 1964)

TABLE 10-10

Illumination Coefficients for Light from Ground

| Window ratio | Illumination coefficient | | | | | |
| | Average (K_ave) | | Minimum (K_min) | | Maximum (K_max) | |
	70%	30%	70%	30%	70%	30%
0.50	0.32	0.18	0.30	0.16	0.33	0.19
1.00	0.41	0.25	0.32	0.20	0.45	0.32
1.50	0.50	0.35	0.35	0.21	0.62	0.45
2.00	0.57	0.42	0.32	0.20	0.83	0.70
2.50	0.62	0.48	0.30	0.19	1.10	0.95
3.00	0.64	0.52	0.28	0.16	1.36	1.24
3.50	0.66	0.55	0.20	0.14	1.68	1.54
4.00	0.66	0.57	0.15	0.09	1.96	1.84
4.50	0.66	0.58	0.10	0.05	2.25	2.15
5.00	0.66	0.59	0.08	0.02	2.53	2.43

Note: Ceiling reflectance, 85%; floor reflectance, 30%; wall reflectances, 70% and 30% as listed.

(Reprinted with permission from Kinzey and Sharp, *Environmental Technologies in Architecture,* Englewood Cliffs, N.J.: Prentice-Hall, Inc., 1964)

TABLE 10-11

Average Footcandles on Vertical Surfaces from an Overcast Sky

| Latitude | December 21 | | | March 21, September 21 | | | June 21 | | |
	8 A.M. 4 P.M.	10 A.M. 2 P.M.	Noon	8 A.M. 4 P.M.	10 A.M. 2 P.M.	Noon	8 A.M. 4 P.M.	10 A.M. 2 P.M.	Noon
30°N	200	500	600	450	900	1200	650	1225	2150
34°N	175	400	500	425	850	1050	650	1200	1850
38°N	150	350	450	400	750	900	650	1200	1750
42°N	100	300	400	375	700	850	650	1150	1500
46°N	50	250	350	350	600	800	650	1100	1400

Note: Based on standard time.

(Reprinted with permission from Kinzey and Sharp, *Environmental Technologies in Architecture,* Englewood Cliffs, N.J.: Prentice-Hall, Inc., 1964)

TABLE 10-12

Average Footcandles on Vertical Surfaces from Clear Sky Only

Latitude	December 21			March 21, September 21			June 21		
	8 A.M. 4 P.M.	10 A.M. 2 P.M.	Noon	8 A.M. 4 P.M.	10 A.M. 2 P.M.	Noon	8 A.M. 4 P.M.	10 A.M. 2 P.M.	Noon
North exposure									
30°N	225	300	300	450	500	525	775	700	500
34°N	175	275	275	400	400	450	675	700	475
38°N	150	275	275	375	400	450	675	650	475
42°N	125	250	250	350	375	400	650	650	475
46°N	75	225	250	350	375	375	650	625	475
South exposure									
30°N	550	950	1125	850	1150	1400	600	800	1200
34°N	555	975	1100	850	1325	1450	675	825	1150
38°N	450	1150	1100	850	1350	1475	675	825	1150
42°N	300	1050	1075	850	1350	1475	675	1000	1250
46°N	200	950	1050	850	1350	1400	675	1050	1350
East exposure									
30°N	775	750	500	1000	1250	750	1400	1325	700
34°N	675	700	475	1200	1300	800	1400	1350	725
38°N	600	650	450	1250	1300	750	1400	1350	700
42°N	375	550	400	1200	1200	725	1450	1300	700
46°N	250	550	400	1150	1075	700	1425	1300	700
West exposure									
30°N	200	350	500	350	450	750	350	500	700
34°N	200	350	475	325	450	800	350	500	700
38°N	175	325	450	300	450	750	350	500	700
42°N	150	300	425	300	400	725	350	500	700
46°N	75	250	400	300	350	700	350	500	700

Note: Based on standard time.

(Reprinted with permission from Kinzey and Sharp, *Environmental Technologies in Architecture*, Englewood Cliffs, N.J.: Prentice-Hall, Inc., 1964)

TABLE 10-13

Average Footcandles on Vertical Surfaces from Sun Only

| | Illumination (fc) on vertical plane facing sun[a] | | | | | | | | |
| | December 21 | | | March 21, September 21 | | | June 21 | | |
Latitude	8 A.M. 4 P.M.	10 A.M. 2 P.M.	Noon	8 A.M. 4 P.M.	10 A.M. 2 P.M.	Noon	8 A.M. 4 P.M.	10 A.M. 2 P.M.	Noon
30°N	4100	6120	6150	5750	5450	4300	6160	3900	1088
34°N	3060	5850	5950	5700	5630	4700	6070	4045	1700
38°N	2480	5530	5875	5620	5850	5110	6070	4125	2310
42°N	2000	5210	5800	5560	5970	5480	6070	4200	2870
46°N	500	4350	5410	5450	6080	5810	6070	4420	3440

[a]For other orientations, multiply by cosine ψ (Figure 10-6).

Note: Based on standard time.

(Reprinted with permission from Kinzey and Sharp, *Environmental Technologies in Architecture*, Englewood Cliffs, N.J.: Prentice-Hall, Inc., 1964)

TABLE 10-14

Average Footcandles on Vertical Surfaces from Ground on a Cloudy Day[a]

| | December 21 | | | March 21, September 21 | | | June 21 | | |
Latitude	8 A.M. 4 P.M.	10 A.M. 2 P.M.	Noon	8 A.M. 4 P.M.	10 A.M. 2 P.M.	Noon	8 A.M. 4 P.M.	10 A.M. 2 P.M.	Noon
30°N	20	50	60	45	90	120	65	123	215
34°N	18	40	50	42	85	100	65	120	185
38°N	15	35	45	40	75	90	65	120	175
42°N	10	30	40	37	70	85	65	115	150
46°N	5	25	35	35	60	80	65	110	140

[a]Average reflection factor of the ground is assumed at 10%.

Note: Based on standard time.

(Reprinted with permission from Kinzey and Sharp, *Environmental Technologies in Architecture*, Englewood Cliffs, N.J.: Prentice-Hall, Inc., 1964)

TABLE 10-15

Average Footcandles on Vertical Surfaces from Ground on a Clear Day

Latitude	December 21			March 21, September 21			June 21		
	8 A.M. 4 P.M.	10 A.M. 2 P.M.	Noon	8 A.M. 4 P.M.	10 A.M. 2 P.M.	Noon	8 A.M. 4 P.M.	10 A.M. 2 P.M.	Noon
North exposure									
30°N	60	200	250	165	350	400	300	430	475
34°N	35	160	210	160	320	380	300	425	465
38°N	25	130	180	140	310	355	280	415	460
42°N	17	105	160	135	275	330	300	405	442
46°N	7	72	115	125	255	312	285	385	415
South exposure									
30°N	90	265	330	215	410	490	280	440	545
34°N	75	230	295	205	410	480	300	435	535
38°N	55	215	260	190	405	460	285	435	530
42°N	35	185	240	185	375	435	300	440	520
46°N	20	145	185	175	355	415	280	440	505
East exposure									
30°N	112	245	270	230	420	425	360	490	495
34°N	85	205	232	240	410	415	375	490	492
38°N	70	165	195	230	400	385	360	485	485
42°N	40	140	175	220	360	360	380	470	465
46°N	25	105	130	205	325	345	360	465	440
West exposure									
30°N	55	205	270	165	340	425	255	410	495
34°N	40	170	232	150	325	415	270	405	490
38°N	28	130	195	135	315	385	255	400	485
42°N	20	110	178	130	280	360	270	390	465
46°N	7	75	130	120	255	345	255	335	440

[a]Average reflection factor of ground is assumed at 10%.

Note: Based on standard time.

(Reprinted with permission from Kinzey and Sharp, *Environmental Technologies in Architecture*, Englewood Cliffs, N.J.: Prentice-Hall, Inc., 1964)

TABLE 10-16

Angular Difference Between Window
Orientation and South

Window facing	Angle (α_2) between perpendicular to window and south (Figure 10-6)
W	90
W × S	79
WSW	68
SW × W	56
SW	45
SW × S	34
SSW	23
S × W	11
S	0
S × E	11
SSE	23
SE × S	34
SE	45
SE × E	56
ESE	68
E × S	79
E	90
E × N	101
ENE	112
NE × E	124
NE	135
NE × N	147
NNE	158
N × E	169
N	180
N × W	169
NNW	158
NW × N	147
NW	135
NW × W	124
WNW	112
W × N	101

(Reprinted with permission from Kinzey and Sharp, *Environmental Technologies in Architecture,* Englewood Cliffs, N.J.: Prentice-Hall, Inc., 1964)

example problems

EXAMPLE PROBLEM 1: DAYLIGHTING (IES PROCEDURE)[2]

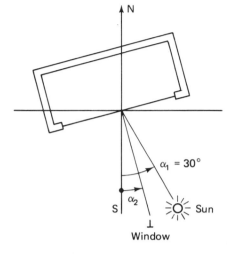

Room 16' x 25'

$8\frac{1}{2}'$ ceiling height

Window - 22' long x 6' high

w/2 ft. sill height

Latitude - 38°N

December 21, 10 a.m.

Determine the illumination contribution on the surface of the window.

1. Find altitude and azimuth, Table 10-2.

$$\text{altitude} = 22° \quad \text{azimuth} = 30°$$

2. Find the contribution from the sun (Table 10-3) for December 21, 10 A.M., 38°N.

$E_{sun\,1} = 6000$ fc,   on a plane normal to a line from the sun

$\qquad = E_1 \times \cos \phi \times \cos (\alpha_1 - \alpha_2)$

solar altitude $\phi = 23°$

solar azimuth $\alpha_1 = 30°$

azimuth to window $\alpha_2 = 11°$

$E_{sun} = 6000 \times \cos 23° \times \cos (30° - 11°)$

$\qquad = 5222$ fc

3. Find the contribution from the sky (Table 10-10). Azimuth from window ψ to sun $= \alpha_1 - \alpha_2$.

$$\text{azimuth} = 19°$$

Interpolate between 0 and 45° for altitude $= 23°$.

$$E_{sky} = 1000 \text{ fc}$$

307

4. Find the contribution from the ground. Assume infinite area.

$$E_{ground} = 1/2 \times \text{ground luminance } (L_G)$$

$$L_G = E_h \times \rho_{ground}$$

$$\rho_{ground} = 6\% \text{ for grass}$$

$$E_h = 2000 \text{ fc} \quad (\text{Table 10-3})$$

$$L_G = 2000 \times 0.06 = 120 \text{ fL}$$

$$E_{ground} = 1/2 \times L_G = 1/2 \times 120 = 60 \text{ fL}$$

5. Total illumination on window.

$$E_{sun} + E_{sky} + E_{ground}$$

$$5222 + 1000 + 60 = 6282$$

6. Determine the illumination at a point in the room (point-by-point method).

General Information

Calculation of illumination at a point from a uniform rectangular source utilizes Figure 10-12, as follows:

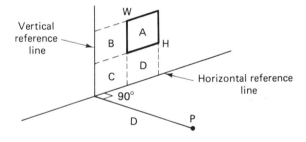

1. Pass a vertical plane through point P to the wall. The intersection of the plane and the wall forms the vertical reference line.

2. All luminous areas must be in contact with both the vertical and horizontal reference lines to use Figure 10-12.

3. To find the illumination at $P(E_p)$ from A, the following values have to be calculated:

$$E_{P_A} = E_{ABCD} - E_{CD} - E_{BC} + E_C$$

Note that the contribution from C was subtracted twice and therefore must be added back once.

Point-by-Point Method for Example Room

Find the illumination at point P at 5 ft in from the window. Note that the actual window is AC, in the following figure.

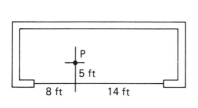 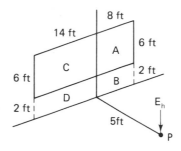

$$EP_{AC} = (E_{AB} - E_B) + (E_{CD} - E_D)$$

From previous calculations, $E_{tot} = 6282$ fc.

$$\text{Window luminance } L_W = E_{tot} \times \tau \times \text{LLF}$$

where τ = transmission of glass
         = 85% (Table 10-4)
   LLF = light-loss factor
         = 0.90 (Table 10-5)
   L_W = 6282 × 0.85 × 0.90 = 4806 ft

From Figure 10-12,

E_{AB}: $D = 5$ ft, $H = 8$ ft, $D/H = 0.63$

         $W = 8$ ft, $H = 8$ ft, $W/H = 1.00$

         $\dfrac{E_h}{L} = 0.100$, $E_{h_{AB}} = 0.100 \times 4806 = 480.6$ fc

E_B: $D = 5$ ft, $H = 2$ ft, $D/H = 2.50$

         $W = 8$ ft, $H = 2$ ft, $W/H = 4.00$

         $\dfrac{E_h}{L} = 0.016$, $E_{h_B} = 0.016 \times 4806 = 76.9$ fc

E_{CD}: $D = 5$ ft, $H = 8$ ft, $D/H = 0.63$

         $W = 14$ ft, $H = 8$ ft, $W/H = 1.75$

         $\dfrac{E_h}{L} = 0.11$ $E_{h_{CD}} = 0.11 \times 4806 = 528.7$ fc

$$E_D: \quad D = 5 \text{ ft}, \quad H = 2 \text{ ft}, \quad D/H = 2.50$$

$$W = 14 \text{ ft}, \quad H = 2 \text{ ft}, \quad W/H = 7.00$$

$$\frac{E_h}{L} = 0.017 \quad E_{h_D} = 0.017 \times 4806 = 81.7 \text{ fc}$$

$$E_p = (E_{h_{AB}} - E_{h_B}) + (E_{h_{CD}} - E_{h_D})$$

$$= (480.6 - 76.9) + (528.7 - 81.7)$$

$$= 850.7 \text{ fc at 5 ft from the window}$$

This assumes that the window is a uniform diffuse surface of 4806 fL. The actual illumination will be higher if direct sunlight reaches point P. This can be determined by making a projection of the window on the floor surface for the specific altitude and azimuth of the sun.

EXAMPLE PROBLEM 2: DAYLIGHTING (KINZEY AND SHARP)[12]

A room 16×25 ft with a $8\frac{1}{2}$ ft ceiling faces south by east; the latitude is 38 °N. Windows extend for 22 ft along the long wall (south side) and are 6 ft high. Glass transmission is 85 percent. The sill is 2 ft above the floor. Net light-admitting area is 80 percent of the structural opening. Compute the average, minimum, and maximum values of illumination under the following conditions: sunny day, 10 A.M. December 21, **no** curtains drawn, 70 percent walls.

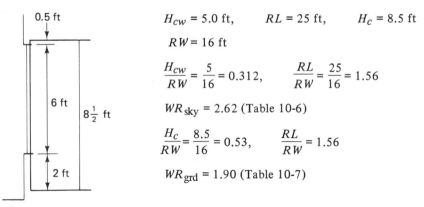

$$H_{cw} = 5.0 \text{ ft}, \quad RL = 25 \text{ ft}, \quad H_c = 8.5 \text{ ft}$$

$$RW = 16 \text{ ft}$$

$$\frac{H_{cw}}{RW} = \frac{5}{16} = 0.312, \qquad \frac{RL}{RW} = \frac{25}{16} = 1.56$$

$$WR_{sky} = 2.62 \text{ (Table 10-6)}$$

$$\frac{H_c}{RW} = \frac{8.5}{16} = 0.53, \qquad \frac{RL}{RW} = 1.56$$

$$WR_{grd} = 1.90 \text{ (Table 10-7)}$$

Illumination Coefficients

Clear Sky and Sun (Table 10-8)

$$K_{ave} = 0.84, \qquad K_{min} = 0.47, \qquad K_{max} = 1.70$$

Ground (Table 10-10)

$$K_{ave} = 0.57, \qquad K_{min} = 0.32, \qquad K_{max} = 0.83$$

Illumination (Tables 10-12, 10-13, 10-15)

$$\text{Clear Sky:} \quad E_v = 1150$$

$$\text{Sun:} \quad E_v = 5530 \text{ (must correct)}$$

$$\text{Ground:} \quad E_v = 215$$

Correct E_v of sun for angle S × E:

$$S \times E, \qquad \alpha_2 = 11° \text{ (Table 10-16)}$$

$$\text{sun azimuth} = 30° \text{ (Table 10-2)}$$

$$\psi = 30° - 11° = 19° \text{ (Figure 10-6)}$$

$$\cos 19° = 0.94$$

$$E_v = 5530 \times 0.94 = 5230 \text{ fc}$$

$$A_g = 22 \times 6 = 132 \times 80\% = 107 \text{ ft}^2$$

Clear Sky

$$E_{ave} = \frac{E_v \times A_g \times K_{ave} \times T_g \times LLF}{A_f} = \frac{1150 \times 107 \times 0.84 \times 0.85 \times 0.90}{16 \times 25}$$

$$= 198 \text{ fc}$$

$$E_{max} = E_{ave} \times \frac{K_{max}}{K_{ave}} = 198 \frac{1.70}{0.84} = 401 \text{ fc}$$

$$E_{min} = 198 \frac{0.47}{0.84} = 111 \text{ fc}$$

Ground

$$E_{ave} = \frac{215 \times 107 \times 0.57 \times 0.85 \times 0.90}{400} = 25 \text{ fc}$$

$$E_{max} = 25 \frac{0.83}{0.57} = 36 \text{ fc}$$

$$E_{min} = 25 \frac{0.32}{0.57} = 14 \text{ fc}$$

Sun

$$E_{\text{ave}} = \frac{5230 \times 107 \times 0.84 \times 0.85 \times 0.90}{400} = 899 \text{ fc}$$

$$E_{\text{max}} = 899 \, \frac{1.70}{0.84} = 1819 \text{ fc}$$

$$E_{\text{min}} = 899 \, \frac{0.47}{0.84} = 503 \text{ fc}$$

	Max	Ave	Min
Sky	401	198	111
Ground	36	25	14
Sun	1819	899	503
	2256	1122	628

index

TABLE 7-6

LDD Graphs (Courtesy of the Illuminating Engineering Society of North America)

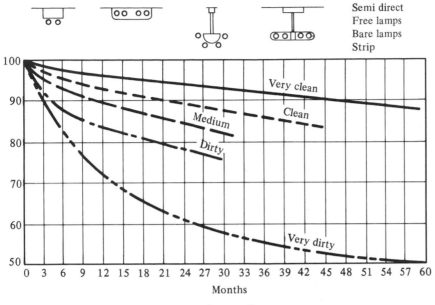

Semi direct
Free lamps
Bare lamps
Strip

Category I

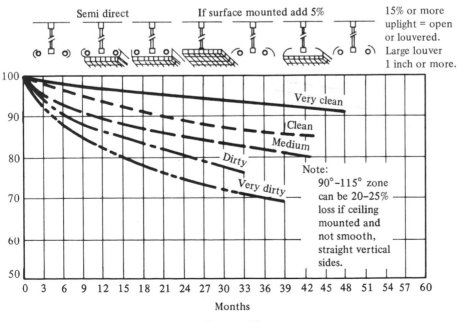

Semi direct   If surface mounted add 5%   15% or more uplight = open or louvered. Large louver 1 inch or more.

Note:
90°–115° zone
can be 20–25%
loss if ceiling
mounted and
not smooth,
straight vertical
sides.

Category II

185

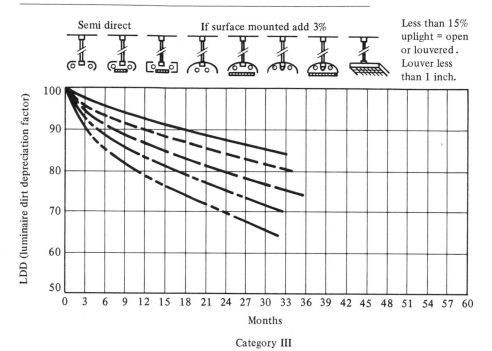

TABLE 7-6 (cont.)

LDD Graphs

Semi direct   If surface mounted add 3%   Less than 15% uplight = open or louvered. Louver less than 1 inch.

LDD (luminaire dirt depreciation factor)

Months

Category **III**

Direct
Closed top recessed
Surface Suspended
Open louvered
Lighted ceiling louvered

Months

186   Category **IV**